D0152985

RWBB
ROGER WILLIAMS COLLEGE LIBRARY
QA431 .M426 1987
Mickens, Ronald E.,
3 1931 00073 6149

DIFFERENCE EQUATIONS

DIFFERENCE EQUATIONS

RONALD E. MICKENS

Callaway Professor of Physics
Atlanta University

VAN NOSTRAND REINHOLD COMPANY
New York

Library of Congress Catalog Card Number 86–10990
ISBN 0–442–26076–8

Printed in the United States of America

Van Nostrand Reinhold Company Inc.
115 Fifth Avenue
New York, New York 10003

Van Nostrand Reinhold Company Limited
Molly Millars Lane
Wokingham, Berkshire RG11 2PY, England

Van Nostrand Reinhold
480 La Trobe Street
Melbourne, Victoria 3000, Australia

Macmillan of Canada
Division of Canada Publishing Corporation
164 Commander Boulevard
Agincourt, Ontario M1S 3C7, Canada

16 15 14 13 12 11 10 9 8 7 6 5 4 3 2 1

Library of Congress Cataloging in Publication Data
Mickens, Ronald E., 1943–
Difference equations.

Bibliography: p.
Includes index.
1. Difference equations. I. Title.
QA431.M426 1987 515′.625 86–10990
ISBN 0–442–26076–8

To my wife
Maria,
my son
James Williamson,
my daughter
Leah Marie.

PREFACE

The purpose of this book is to present and explain mathematical methods for determining solutions to linear and nonlinear difference equations.

Within the past two decades the study of difference equations has acquired a new significance. This comes about, in large part, from the use of these equations in the formulation and analysis of discrete-time systems, the numerical integration of differential equations by finite-difference schemes, the study of deterministic chaos, etc. Currently, no modern book exists which provides easy access to the broad range of concepts and solution techniques relating to difference equations. This book fills this gap by providing the necessary background material to help the reader both to understand the research literature of these developments and (hopefully) to contribute useful results to it.

This book will be useful to both advanced undergraduates and more experienced researchers. The minimum prerequisites for successful use of the book are a knowledge of intermediate calculus and a standard first course in differential equations. The Appendix introduces, in summary fashion, a number of mathematical relations and concepts which are needed for an understanding of the material in the text. We advise all readers to review this section before beginning to tackle the book.

A major feature of this book is that, after the introduction of most concepts and techniques, a number of examples are worked in sufficient detail that the reader can gain a complete grasp of the particular concept or technique. Consequently, the book should be very useful for self-study.

While, in general, we present most theorems and major results in rigorous mathematical form, our primary concern is the presentation of the various concepts and solution techniques in such a manner that useful application to problems that give rise to difference equations can readily occur.

In Chapter 1, we begin with a section on the genesis of difference equations. This is followed by sections which give the important properties of difference equations; an existence and uniqueness theorem; the introduction and use of the difference, Δ, and shift, E, operators; and the factorial polynomials. The chapter ends with a discussion of the Δ^{-1} operator and the sum calculus.

Chapter 2 is concerned with a general discussion of first-order difference equations. In addition to the general linear equation, a number of special cases are treated in detail. The chapter concludes with two sections on nonlinear equations: a graphic geometrical method and an analytic expansion technique are presented and used to obtain information on the asymptotic behavior of possible solutions.

Chapter 3 introduces the notion of linearly independent functions and the Casoratian determinant. These concepts are used to prove and interpret a number of theorems on the general nth-order linear difference equation. General methods are then presented for solving both homogeneous and inhomogeneous equations. The case of second-order equations is discussed in detail. The last section treats Sturm–Liouville difference equation systems.

In Chapter 4, a complete discussion is given of linear difference equations having constant coefficients. The characteristic function is introduced and, from its properties, it is shown how the general solution to homogeneous difference equations can be determined. For special forms of the inhomogeneous term, we give two methods, undetermined coefficients and operator techniques, for calculating particular solutions. In addition, the z-transform method is defined and used to solve linear difference equations. Section 4.4 gives an interesting and not well-known relationship between linear differential and difference equations.

Chapter 5 is concerned with linear partial difference equations having constant coefficients. Solutions to homogeneous equations are obtained by using symbolic, Lagrange's, separation of variables, and Laplace's methods. Techniques for determining particular solutions for the case of inhomogeneous equations, where the inhomogeneous terms have special forms, are presented. A brief discussion of simultaneous partial difference equations is also given.

The final chapter presents a variety of special forms of nonlinear ordinary and partial difference equations for which exact solutions can be determined. The basic technique is to find a nonlinear transformation which converts a given nonlinear equation into a linear equation that can be solved.

References to certain topics presented in the book are given in the Notes and References section. These references allow the reader to see a given topic discussed at a more advanced and rigorous level. In certain instances, the references are to "nonstandard" topics such as that in Section 4.4 on the relation between linear difference and differential equations.

The Bibliography contains books that either are in my personal library or were accessible to me during the writing of this book. It is separated into two sections: the first lists books having a general theoretical orientation, while the second section gives books that apply difference equations to a variety of problems in the sciences and mathematics.

I thank my many colleagues and students for their interest and suggestions on this book. I am particularly grateful to Ms. Annette Rohrs for her efforts in typing the manuscript. I also wish to thank the Department of Energy, the National Aeronautics and Space Administration, and the UNCF Distinguished Fellowship Program for support of my research efforts and for providing the funds for acquiring release time to write this book.

Atlanta, Georgia

Ronald E. Mickens

CONTENTS

1
THE DIFFERENCE CALCULUS

1.1. GENESIS OF DIFFERENCE EQUATIONS

We begin this chapter with a number of examples which illustrate how difference equations arise. These examples also show the diversity of the areas in which difference equations apply. In general, we expect difference equations to occur whenever the system under study depends on one or more variables that can only assume a discrete set of possible values.

1.1.1. Example A

The Chebyshev polynomials are defined by the expression

$$C_k(x) = \frac{1}{2^{k-1}} \cos(k \cos^{-1} x), \qquad k = 0, 1, 2, \ldots, |x| < 1. \tag{1.1}$$

We now show that these functions obey the following recurrence relation:

$$C_{k+1}(x) - xC_k(x) + (1/4)\, C_{k-1}(x) = 0, \tag{1.2}$$

where, from equation (1.1), we have

$$C_0(x) = 2, \qquad C_1(x) = x. \tag{1.3}$$

From the fact that

$$\cos(\theta_1 \pm \theta_2) = \cos\theta_1 \cos\theta_2 \mp \sin\theta_1 \sin\theta_2, \tag{1.4}$$

it follows that

$$\begin{aligned} C_{k+1}(x) &= \frac{1}{2^k} \cos[(k+1)\cos^{-1} x] \\ &= \frac{1}{2^k} \cos(\cos^{-1} x)\cos(k \cos^{-1} x) \\ &\quad - \frac{1}{2^k} \sin(\cos^{-1} x)\sin(k \cos^{-1} x), \end{aligned} \tag{1.5}$$

and

$$C_{k-1}(x) = \frac{1}{2^{k-2}} \cos[(k-1)\cos^{-1} x]$$
$$= \frac{4}{2^k} \cos(\cos^{-1} x)\cos(k \cos^{-1} x) \qquad (1.6)$$
$$+ \frac{4}{2^k} \sin(\cos^{-1} x)\sin(k \cos^{-1} x).$$

Therefore,

$$C_{k+1}(x) + (1/4)\, C_{k-1}(x) = \frac{2}{2^k} \cos(\cos^{-1} x)\cos(k \cos^{-1} x)$$
$$= \frac{x}{2^{k-1}} \cos(k \cos^{-1} x) \qquad (1.7)$$
$$= xC_k(x),$$

which is just equation (1.2).

1.1.2. Example B

Consider the integral for non-negative integer k:

$$I_k(\phi) = \int_0^{\pi} \frac{\cos(k\theta) - \cos(k\phi)}{\cos\theta - \cos\phi}\, d\theta, \qquad (1.8)$$

where $I_0 = 0$ and $I_1 = \pi$. (Note, there is no singularity at $\theta = \phi$ since the integrand at these values is equal to k^2 by application of the appropriate limit theorem.) Since I_k is known for $k = 0$ and $k = 1$, let us try to determine a linear relation among I_{k+1}, I_k, and I_{k-1}. If such a relation can be found, then it can be used to calculate I_k recursively.

Define the operator L as follows:

$$LI_k = AI_{k+1} + BI_k + CI_{k-1}, \qquad (1.9)$$

where A, B, and C are constants independent of k and θ; however, they may depend on ϕ. Therefore,

$$L \cos(k\theta) = A\cos(k+1)\theta + B\cos(k\theta) + C\cos(k-1)\theta$$
$$= [(A+C)\cos\theta + B]\cos(k\theta) - [(A-c)\sin\theta]\sin(k\theta), \qquad (1.10)$$

where use has been made of equation (1.4). If we take $A = C$, $B = -2A \cos\phi$, and set $A = 1$, then equation (1.10) becomes

$$L\cos(k\theta) = 2(\cos\theta - \cos\phi)\cos(k\theta), \tag{1.11}$$

which is proportional to the denominator of the integrand of equation (1.8). Now applying the operator L to $\cos(k\phi)$ gives the result

$$L\cos(k\theta) = 0. \tag{1.12}$$

Therefore, applying L to both sides of equation (1.8) gives

$$\begin{aligned} LI_k(\phi) &= \int_0^\pi \frac{L\cos(k\theta) - L\cos(k\phi)}{\cos\theta - \cos\phi} \\ &= 2\int_0^\pi \cos(k\theta)d\theta = 0. \end{aligned} \tag{1.13}$$

Consequently, we conclude that the integral $I_k(\phi)$ satisfies the relation

$$I_{k+1} - (2\cos\phi)I_k + I_{k-1} = 0 \tag{1.14}$$

for $k = 1, 2, \ldots$. Since I_0 and I_1 are known, I_k can be determined recursively for any positive integer value of k.

1.1.3. Example C

Consider a plane that has lying in it k nonparallel lines. Into how many separate compartments will the plane be divided if not more than two lines intersect in the same point?

Let N_k be the number of compartments. Therefore, the $(k + 1)$th line will be cut by the k previous lines in k points and, consequently, divides each of the $k + 1$ prior existing compartments into two. This gives

$$N_{k+1} = N_k + (k+1). \tag{1.15}$$

Note for $k = 0$, $N_0 = 1$, since the plane is then undivided. For $k = 1$, $N_1 = 2$, because a single line divides the plane into two compartments, etc.

1.1.4. Example D

Let us now consider a single population whose size is known at a discrete set of time intervals, $t_k = (\Delta t)k$, where Δt is a fixed interval and k is a non-negative integer. Denote the size of the population at time t_k by P_k.

Now assume that when the population is small, the change in the population from time t_k to t_{k+1} is given by the expression

$$P_{k+1} = \alpha P_k, \tag{1.16}$$

where α is a positive constant. Further assume that at large population sizes, there is competition among the members of the population (for food and other resources). This competition can be modeled by adding to equation (1.16) a term of the form $-\beta P_k(P_k - 1)$, where $\beta > 0$. Roughly, this means that the magnitude of the competition is proportional to the total number of possible interactions among the members of the population and that competition has a negative effect on the growth of the population. Combining these two results gives the following population dynamics equation:

$$P_{k+1} = \alpha P_k - \beta P_k(P_k - 1). \tag{1.17}$$

This equation can be written to give

$$P_{k+1} = (\alpha + \beta)P_k - \beta P_k^2. \tag{1.18}$$

The coefficient $\alpha + \beta$ is the effective or net "birth rate," while β is a measure of the competition. Note that α is the birth rate in the absence of competition. From this simple model, we can conclude that when competition occurs the net birth rate increases.

1.1.5. Example E

Suppose the following differential equation:

$$\frac{dy}{dt} = f(y, t), \tag{1.19}$$

where $f(y, t)$ is a given function of y and t, cannot be integrated in closed form in terms of the elementary functions. We could proceed by using the following simple scheme to determine a numerical solution.

First, construct a lattice, $t_k = (\Delta t)k$, where Δt is a fixed t interval and k is the set of integers. Second, replace the derivative by the approximation,

$$\frac{dy(t)}{dt} \rightarrow \frac{y(t + \Delta t) - y(t)}{\Delta t} = \frac{y_{k+1} - y_k}{\Delta t}, \tag{1.20}$$

where y_k is the approximation to the exact solution of equation (1.19) at $t = t_k$, i.e.,

$$y_k \simeq y(t_k). \tag{1.21}$$

Finally, replace the right-hand side of equation (1.19) by

$$f(y, t) \rightarrow f[y_k, (\Delta t)k]. \tag{1.22}$$

Putting all of this together gives

$$\frac{y_{k+1} - y_k}{\Delta t} = f[y_k, (\Delta t)k], \tag{1.23}$$

or

$$y_{k+1} = y_k + (\Delta t) f[y_k, (\Delta t)k]. \tag{1.24}$$

If y_0 is specified, then y_k for $k = 1, 2, \ldots,$ can be determined.

1.1.6. Example F

Let us determine a power-series solution

$$y(x) = \sum_{k=0}^{\infty} C_k x^k \tag{1.25}$$

to the differential equation

$$\frac{d^2y}{dx^2} + 3x\frac{dy}{dx} + 3y = 0, \tag{1.26}$$

where the coefficients, C_k, are to be found. We have

$$x\frac{dy}{dx} = x\sum_{k=0}^{\infty} kC_k x^{k-1} = \sum_{k=2}^{\infty} (k-2)C_{k-2}x^{k-2}, \tag{1.27}$$

$$\frac{d^2y}{dx^2} = \sum_{k=0}^{\infty} k(k-1)C_k x^{k-2} = \sum_{k=2}^{\infty} k(k-1)C_k x^{k-2}, \tag{1.28}$$

and

$$y = \sum_{k=0}^{\infty} C_k x^k = \sum_{k=2}^{\infty} C_{k-2} x^{k-2}. \tag{1.29}$$

Substitution of equations (1.27), (1.28), and (1.29) into equation (1.26) gives

$$\sum_{k=2}^{\infty} [k(k-1)C_k + 3(k-1)C_{k-2}]x^{k-2} = 0. \tag{1.30}$$

Equating each coefficient to zero gives the following recursion relation which the C_k must satisfy:

$$kC_k + 3C_{k-2} = 0. \tag{1.31}$$

The solution of this recursion relation, when substituted into equation (1.25), provides the required power series solution to equation (1.26).

1.2. DEFINITIONS

A sequence is a function whose domain is the set of integers. In the main, we will consider sequences whose domains are the non-negative integers. However, this restriction is of no importance; if the sequence has some other domain, it will be clear from the context what that domain is. We denote the general member of a sequence by y_k and use the notation $\{y_k\}$ to represent the sequence $y_0, y_1, y_2, \ldots$.

Now consider a given sequence $\{y_k\}$ and let there be a rule for determining the general member y_k. Assume that this rule or relation takes the form

$$y_{k+n} = F(k, y_{k+n-1}, y_{k+n-2}, \ldots, y_k), \tag{1.32}$$

where F is a well-defined function of its arguments. Thus, given appropriate starting values, all the remaining members of the sequence can be generated. In fact, from equation (1.32), it is clear that if n successive values of y_k are specified, then the sequence $\{y_k\}$ is uniquely defined. These specified values are called initial conditions.

The following definition shows the connection between sequences and difference equations.

Definition. An ordinary difference equation is a relation of the form given by equation (1.32).

Several additional concepts need to be introduced. They are provided in the following definitions.

Definition. The order of a difference equation is the difference between the highest and lowest indices that appear in the equation.

The expression given by equation (1.32) is an nth-order difference equation if and only if the term y_k appears in the function F on the right-hand side. Note that shifts in the labeling of the indices do not change the order of a difference equation. For example,

$$y_{k+n+r} = F(k + r, y_{k+n+r-1}, y_{k+n+r-2}, \ldots, y_{k+r}) \qquad (1.33)$$

is the nth-order difference equation which is equivalent to equation (1.32).

Definition. A difference equation is linear if it can be put in the form

$$y_{k+n} + a_1(k)y_{k+n-1} + a_2(k)y_{k+n-2} + \cdots + a_{n-1}(k)y_{k+1} + a_n(k)y_k = R_k, \qquad (1.34)$$

where $a_i(k)$, $i = 1, 2, \ldots, n$, and R_k are given functions of k.

Definition. A difference equation is nonlinear if it is not linear.

Definition. A solution of a difference equation is a function $\phi(k)$ that reduces the equation to an identity.

The following examples highlight the concepts introduced in the above definitions.

1.2.1. Example A

We indicate for each of the following difference equations its order and whether it is linear or not.

$$y_{k+1} - 3y_k + y_{k-1} = e^{-k} \qquad \text{(second order, linear)}, \qquad (1.35a)$$

$$y_{k+1} = y_k^2 \qquad \text{(first order, nonlinear)}, \qquad (1.35b)$$

$$y_{k+4} - y_k = k2^k \qquad \text{(fourth order, linear)}, \qquad (1.35c)$$

$$y_{k+1} = y_k - (1/100)y_k^2 \qquad \text{(first order, nonlinear)}, \qquad (1.35d)$$

$$y_{k+3} = \cos y_k \qquad \text{(third order, nonlinear)}, \qquad (1.35e)$$

$$y_{k+2} + (3k - 1)y_{k+1} - \frac{k}{k+1} y_k = 0 \qquad \text{(second order, linear)}. \qquad (1.35f)$$

1.2.2. Example B

The function

$$\phi(k) = 2^k \tag{1.36}$$

is a solution of the first-order linear difference equation

$$y_{k+1} - 2y_k = 0, \tag{1.37}$$

since on substitution of $\phi(k)$ into the equation an identity is obtained in the form

$$2^{k+1} - 2 \cdot 2^k = 0. \tag{1.38}$$

1.2.3 Example C

The first-order nonlinear equation

$$y_{k+1}^2 - y_k^2 = 1 \tag{1.39}$$

has the solution

$$\phi(k) = \sqrt{k + c}, \tag{1.40}$$

where c is a constant. This statement can be checked by substituting $\phi(k)$ into the left-hand side of equation (1.39) to obtain

$$(\sqrt{k + 1 + c})^2 - (\sqrt{k + c})^2 = (k + 1 + c) - (k + c) = 1, \tag{1.41}$$

which is equal to the expression on the right-hand side.

1.2.4. Example D

The second-order linear equation

$$y_{k+1} - y_{k-1} = 0 \tag{1.42}$$

has two solutions,

$$\phi_1(k) = (-1)^k, \qquad \phi_2(k) = 1, \tag{1.43}$$

as can be easily shown by direct substitution into equation (1.42).

Let c_1 and c_2 be two arbitrary constants. We now show that

$$\phi(k) = c_1\phi_1(k) + c_2\phi_2(k) = c_1(-1)^k + c_2 \tag{1.44}$$

is also a solution. Substitution of this expression into the left-hand side of equation (1.42) gives

$$c_1(-1)^{k+1} + c_2 - c_1(-1)^{k-1} - c_2 = 0. \tag{1.45}$$

1.3. DERIVATION OF DIFFERENCE EQUATIONS

Assume that the general member y_k, of a sequence $\{y_k\}$, is defined in terms of a specified function of k and n arbitary constants $c_1, c_2, \ldots, c_n$. We now show that y_k satisfies an nth-order difference equation.

By assumption, we have

$$y_k = f(k, c_1, c_2, \ldots, c_n) \tag{1.46}$$

and

$$\begin{aligned} y_{k+1} &= f(k+1, c_1, c_2, \ldots, c_n), \\ &\vdots \\ y_{k+n} &= f(k+n, c_1, c_2, \ldots, c_n). \end{aligned} \tag{1.47}$$

These are a set of $n+1$ equations. If now the n constants c_i are eliminated, then a relation of the form

$$G(k, y_k, y_{k+1}, \ldots, y_{k+n}) = 0 \tag{1.48}$$

is obtained. This is an nth-order difference equation. We conclude that if the general member y_k of a sequence $\{y_k\}$ is expressed as a function of k and n arbitrary constants, then y_k satisfies an nth-order difference equation.

1.3.1. Example A

Let the general member y_k, of the sequence $\{y_k\}$, be defined as

$$y_k = A2^k, \tag{1.49}$$

where A is an arbitrary constant. Since there is one arbitrary constant, the difference equation whose solution is given by equation (1.49) will be of first order. It can be found as follows:

$$y_{k+1} = A2^{k+1} = 2A2^k = 2y_k. \tag{1.50}$$

1.3.2. Example B

Let y_k be given by the expression

$$y_k = c_1 2^k + c_2 5^k, \tag{1.51}$$

where c_1 and c_2 are arbitrary constants. Thus, y_k must satisfy a second-order difference equation. To determine this equation, we calculate y_{k+1} and y_{k+2}:

$$\begin{aligned} y_{k+1} &= 2c_1 2^k + 5c_2 5^k, \\ y_{k+2} &= 4c_1 2^k + 25c_2 5^k. \end{aligned} \tag{1.52}$$

The elimination of c_1 and c_2 gives the relation

$$\begin{vmatrix} y_k & 1 & 1 \\ y_{k+1} & 2 & 5 \\ y_{k+2} & 4 & 25 \end{vmatrix} = 0, \tag{1.53}$$

which, on expanding, is equal to

$$y_{k+2} - 7y_{k+1} + 10y_k = 0. \tag{1.54}$$

1.3.3. Example C

Consider y_k defined as

$$y_k = Ak + f(A), \tag{1.55}$$

where A is an arbitrary constant and f can be an arbitrary function. Now

$$y_{k+1} = Ak + f(A) + A = y_k + A, \tag{1.56}$$

or

$$A = y_{k+1} - y_k. \tag{1.57}$$

Substitution of the result of equation (1.57) into equation (1.55) gives

$$y_k = (y_{k+1} - y_k)k + f(y_{k+1} - y_k) \tag{1.58}$$

as the general nonlinear first-order difference equation for which the y_k of equation (1.55) is a solution.

1.3.4. Example D

The function

$$y_k = (c_1 + c_2 k)2^k, \tag{1.59}$$

where c_1 and c_2 are arbitrary constants, is the solution to a second-order difference equation. To determine this equation, we first calculate y_{k+1} and y_{k+2}:

$$y_{k+1} = 2c_1 2^k + 2c_2(k+1)2^k, \tag{1.60}$$

$$y_{k+2} = 4c_1 2^k + 4c_2(k+2)2^k. \tag{1.61}$$

Multiplying equation (1.60) by 2 and substracting this from equation (1.61) gives an expression that can be solved for c_2:

$$c_2 = \frac{1}{2^{k+2}}(y_{k+2} - 2y_{k+1}), \tag{1.62}$$

or

$$c_2 k 2^k = (k/4)(y_{k+2} - 2y_{k+1}). \tag{1.63}$$

Substituting this last result into equation (1.60) gives

$$c_1 2^k = \tfrac{1}{2} y_{k+1} - \tfrac{1}{4}(k+1)(y_{k+2} - 2y_{k+1}). \tag{1.64}$$

If equations (1.63) and (1.64) are used in the right-hand side of equation (1.59) and if the resulting expression is simplified, then the following result is obtained:

$$y_{k+2} - 4y_{k+1} + 4y_k = 0. \tag{1.65}$$

This is the difference equation whose solution is equation (1.59).

1.4. EXISTENCE AND UNIQUENESS THEOREM

It should be clear that for a given difference equation, even if a solution is known to exist, there is no assurance that it will be unique. The solution must be further restricted by giving a set of initial conditions equal in number to the order of the equation. The following theorem states conditions which assure the existence of a unique solution.

Theorem 1.1: Let

$$y_{k+n} = f(k, y_k, y_{k+1}, \ldots, y_{k+n-1}), \qquad k = 0, 1, 2, 3, \ldots, \tag{1.66}$$

be an nth-order difference equation where f is defined for each of its arguments. This equation has one and only one solution corresponding to each arbitrary selection of the n initial values $y_0, y_1, \ldots, y_{n-1}$.

Proof: If the values $y_0, y_1, \ldots, y_{n-1}$ are given, then the difference equation with $k = 0$ uniquely specifies y_n. Once y_n is known, the difference equation with $k = 1$ gives y_{n+1}. Proceeding in this way, all y_k, for $k \geq n$, can be determined.

1.5. OPERATORS Δ AND *E*

The operator Δ is defined as follows:

$$\Delta y_k \equiv y_{k+1} - y_k. \tag{1.67}$$

The expression $y_{k+1} - y_k$ is called the difference of y_k; correspondingly, we call Δ the (first) difference operator.

The second difference operator is denoted $\Delta^2 = \Delta \cdot \Delta$ and when acting on y_k produces the result

$$\begin{aligned} \Delta^2 y_k &= \Delta(\Delta y_k) = \Delta(y_{k+1} - y_k) \\ &= \Delta y_{k+1} - \Delta y_k = (y_{k+2} - y_{k+1}) - (y_{k+1} - y_k) \\ &= y_{k+2} - 2y_{k+1} + y_k. \end{aligned} \tag{1.68}$$

In general, for positive n, we define the relation

$$\Delta(\Delta^n y_k) = \Delta^{n+1} y_k. \tag{1.69}$$

It therefore should be clear that

$$\Delta^m \Delta^n y_k = \Delta^n \Delta^m y_k = \Delta^{m+n} y_k, \tag{1.70}$$

for m and n positive integers. Assuming equation (1.70) to hold true when $m = 0$, we find

$$\Delta^0 \Delta^n y_k = \Delta^{n+0} y_k = \Delta^n \Delta^0 y_k, \tag{1.71}$$

which implies that Δ^0 is the identity operator

$$\Delta^0 y_k = y_k. \tag{1.72}$$

It should be obvious that Δ satisfies both the distributive law

$$\Delta(x_k + y_k) = \Delta x_k + \Delta y_k \tag{1.73}$$

and the commutative law

$$\Delta(c y_k) = c\,\Delta y_k, \tag{1.74}$$

where c is a constant. Therefore, Δ is a linear operator

$$\Delta(c_1 x_k + c_2 y_k) = c_1 \Delta x_k + c\,\Delta y_k, \tag{1.75}$$

where c_1 and c_2 are constants.

Theorem 1.2:

$$\begin{aligned} \Delta^n y_k = y_{k+n} - n y_{k+n-1} &+ \frac{n(n-1)}{2!} y_{k+n-2} \\ &+ \cdots + (-1)^i \frac{n(n-1)\cdots(n-i+1)}{i!} y_{k+n-i} \\ &+ \cdots + (-1)^n y_k. \end{aligned} \tag{1.76}$$

Proof: Mathematical induction will be used to prove this theorem. The relation is valid for $n = 1$, since in this case we obtain the result of equation (1.67). Now, in equation (1.76) replace k by $k + 1$; doing this gives

$$\begin{aligned} \Delta^n y_{k+1} = y_{k+n+1} - n y_{k+n} &+ \frac{n(n-1)}{2} y_{k+n-1} \\ &+ \cdots + (-1)^{i+1} \frac{n(n-1)\cdots(n-i)}{(i+1)!} y_{k+n-i} \\ &+ \cdots + (-1)^n y_{k+1}. \end{aligned} \tag{1.77}$$

Now

$$\begin{aligned}\Delta(\Delta^n y_k) &= \Delta^n y_{k+1} - \Delta^n y_k \\ &= y_{k+n+1} - (n+1)y_{k+n} + \frac{(n+1)n}{2!} y_{k+n-1} \\ &\quad + \cdots + (-1)^{i+1} \frac{(n+1)n(n-1)\cdots(n-i+1)}{(i+1)!} y_{k+n-i} \\ &\quad + \cdots + (-1)^{n+1} y_k.\end{aligned} \tag{1.78}$$

This last expression is just equation (1.76) with n replaced by $n + 1$. Therefore, by the method of induction the result given by equation (1.76) holds true for all positive integers n.

Using the definition for the binomial coefficients,

$$\binom{n}{i} = \frac{n(n-1)\cdots(n-i+1)}{i!} = \frac{n!}{(i!)(n-i)!}, \tag{1.79}$$

we can rewrite the result of the above theorem to read

$$\Delta^n y_k = \sum_{i=0}^{n} (-1)^i \binom{n}{i} y_{k+n-i}. \tag{1.80}$$

We now consider functions of the operator Δ. Let $f(r)$ be a polynomial in the variable r:

$$f(r) = a_0 r^m + a_1 r^{m-1} + \cdots + a_m, \tag{1.81}$$

where $a_0, a_1, \ldots, a_m$ are constants. The operator function, $f(\Delta)$, is defined as follows:

$$\begin{aligned}f(\Delta)y_k &= (a_0\Delta^m + a_1\Delta^{m-1} + \cdots + a_m)y_k \\ &= a_0\Delta^m y_k + a_1\Delta^{m-1}y_k + \cdots + a_m y_k.\end{aligned} \tag{1.82}$$

Let α_1, α_2, β_1, and β_2 be constants; therefore,

$$\begin{aligned}(\alpha_1 + \beta_1\Delta)(\alpha_2 + \beta_2\Delta)y_k &= \alpha_1(\alpha_2 + \beta_2\Delta)y_k + \beta_1\Delta(\alpha_2 + \beta_2\Delta)y_k \\ &= \alpha_1\alpha_2 y_k + \alpha_1\beta_2\Delta y_k + \beta_1\alpha_2\Delta y_k + \beta_1\beta_2\Delta^2 y_k \\ &= \alpha_1\alpha_2 y_k + (\alpha_1\beta_2 + \alpha_2\beta_1)\Delta y_k + \beta_1\beta_2\Delta^2 y_k.\end{aligned} \tag{1.83}$$

Likewise, an easy calculation shows that

$$(\alpha_2 + \beta_2\Delta)(\alpha_1 + \beta_1\Delta)y_k = \alpha_1\alpha_2 y_k + (\alpha_1\beta_2 + \alpha_2\beta_1)\Delta y_k + \beta_1\beta_2\Delta^2 y_k. \quad (1.84)$$

We conclude that the order of the two operators, $(\alpha_1 + \beta_1\Delta)$ and $(\alpha_2 + \beta_2\Delta)$, is unimportant provided the coefficients α_1, α_2, β_1, and β_2 are not functions of k, i.e., the coefficients are constants. This result can be generalized to an operator having m such linear factors. Thus, since $f(r)$, in equation (1.81), is a polynomial function of degree m, it can be factorized in the form

$$f(r) = (r - r_1)(r - r_2) \cdots (r - r_m) = \prod_{i=1}^{m} (r - r_i), \quad (1.85)$$

and the result of equation (1.82) can be written as

$$f(\Delta)y_k = \prod_{i=1}^{m} (\Delta - r_i)y_k. \quad (1.86)$$

We now introduce a new operator, E, which can act on y_k. For p any positive or negative integer, we have

$$E^p y_k = y_{k+p}. \quad (1.87)$$

Consequently, when E^p operates on a function of k, it gives a new function shifted by p units. It follows from the definition of E that

$$E^0 y_k = y_{k+0} = y_k; \quad (1.88)$$

therefore, E^0 is the identity operator. We call E the shift operator.

The following is a list of some of the elementary properties of the shift operator, E:

$$E^p(c_1 x_k + c_2 y_k) = c_1 x_{k+p} + c_2 y_{k+p}, \quad (1.89)$$

$$E^p E^q y_k = E^q E^p y_k = E^{p+q} y_k, \quad (1.90)$$

$$f(E)y_k = \prod_{i=1}^{m} (E - r_i)y_k. \quad (1.91)$$

In the above relations, c_1 and c_2 are constants, p and q are integers, and $f(r)$ is a polynomial function of degree m.

From the definition of the two operators Δ and E, it follows that they are related. Since $\Delta y_k = y_{k+1} - y_k$ and $Ey_k = y_{k+1}$, we obtain

$$\Delta y_k = (E - 1)y_k. \tag{1.92}$$

Therefore, the operator Δ is equivalent mathematically to the operator $E - 1$, where 1 is the identity operator, i.e., $1 \cdot y_k = y_k$. In symbolic form

$$\Delta \equiv E - 1. \tag{1.93}$$

Also, since $y_{k+1} = y_k + \Delta y_k$, we have

$$E \equiv 1 + \Delta. \tag{1.94}$$

If $f(r)$ and $g(r)$ are polynomial functions of the variable r, then

$$\begin{aligned} f(E) &= f(1 + \Delta), \\ g(\Delta) &= g(E - 1). \end{aligned} \tag{1.95}$$

The above results can be used to prove the following theorem, which is the inverse problem to Theorem 1.2.

Theorem 1.3:

$$\begin{aligned} y_{k+n} &= y_k + n\,\Delta y_k + \frac{n(n-1)}{2!}\Delta^2 y_k \\ &\quad + \cdots + \binom{n}{i}\Delta^i y_k + \cdots + \Delta^n y_k. \end{aligned} \tag{1.96}$$

Proof: Write y_{k+n} as

$$y_{k+n} = E^n y_k = (1 + \Delta)^n y_k = \sum_{i=0}^{n} \binom{n}{i} \Delta^i y_k. \tag{1.97}$$

The last expression is equation (1.96).

1.6. ELEMENTARY DIFFERENCE OPERATORS

Let x_k and y_k be functions of k. We now establish certain results for the difference of a product, difference of a quotient, and difference of a finite sum.

Difference of a Product

Apply the difference operator to the product $x_k y_k$:

$$\begin{aligned} \Delta(x_k y_k) &= x_{k+1} y_{k+1} - x_k y_k \\ &= x_{k+1} y_{k+1} - x_{k+1} y_k + y_k x_{k+1} - x_k y_k \\ &= x_{k+1}(y_{k+1} - y_k) + y_k (x_{k+1} - x_k) \\ &= x_{k+1} \Delta y_k + y_k \Delta x_k. \end{aligned} \tag{1.98}$$

Leibnitz's Theorem for Differences

We now prove that

$$\begin{aligned} \Delta^n(x_k y_k) = x_k \Delta^n y_k &+ \binom{n}{1} (\Delta x_k)(\Delta^{n-1} y_{k+1}) \\ &+ \binom{n}{2} (\Delta^2 x_k)(\Delta^{n-2} y_{k+2}) \\ &+ \cdots + \binom{n}{n} (\Delta^n x_k)(y_{k+n}). \end{aligned} \tag{1.99}$$

Proof: Define operators E_1 and E_2 which operate, respectively, only on x_k and y_k. Therefore,

$$E_1(x_k y_k) = x_{k+1} y_k, \qquad E_2(x_k y_k) = x_k y_{k+1}. \tag{1.100}$$

From

$$E_1 E_2(x_k y_k) = x_{k+1} y_{k+1}, \tag{1.101}$$

we conclude that $E = E_1 E_2$. Define additional operators Δ_1 and Δ_2 such that

$$\Delta_1 = E_1 - 1, \qquad \Delta_2 = E_2 - 1. \tag{1.102}$$

Therefore,

$$\begin{aligned} \Delta &= E - 1 = E_1 E_2 - 1 = (1 + \Delta_1) E_2 - 1 \\ &= E_2 + \Delta_1 E_2 - 1 = \Delta_2 + \Delta_1 E_2 \end{aligned} \tag{1.103}$$

and

$$\Delta^n(x_k y_k) = (\Delta_2 + \Delta_1 E_2)^n (x_k y_k). \tag{1.104}$$

Expanding the term $(\Delta_2 + \Delta_1 E_2)^n$ by the binomial theorem gives

$$\begin{aligned}
\Delta^n(x_k y_k) &= \left[\Delta_2{}^n + \binom{n}{1} \Delta_2{}^{n-1}\Delta_1 E_2 + \binom{n}{2} \Delta_2{}^{n-2}\Delta_1{}^2 E_2{}^2 + \cdots \right. \\
&\quad \left. + \binom{n}{n} \Delta_1{}^n E_2{}^n \right] (x_k y_k) \\
&= x_k \, \Delta^n y_k + \binom{n}{1} (\Delta x_k)(\Delta^{n-1} y_{k+1}) + \cdots \\
&\quad + \binom{n}{n} (\Delta^n x_k)(y_{k+n}),
\end{aligned} \tag{1.105}$$

which is just the result of equation (1.99).

Difference of a Quotient

Applying the difference operator to the quotient x_k/y_k gives

$$\begin{aligned}
\Delta\left(\frac{x_k}{y_k}\right) &= \frac{x_{k+1}}{y_{k+1}} - \frac{x_k}{y_k} = \frac{x_{k+1}y_k - y_{k+1}x_k}{y_k y_{k+1}} \\
&= \frac{(x_{k+1} - x_k)y_k - x_k(y_{k+1} - y_k)}{y_k y_{k+1}} \\
&= \frac{y_k \, \Delta x_k - x_k \, \Delta y_k}{y_k y_{k+1}}.
\end{aligned} \tag{1.106}$$

Difference of a Finite sum

Let

$$S_k = y_1 + y_2 + y_3 + \cdots + y_k. \tag{1.107}$$

Therefore,

$$S_{k+1} = y_1 + y_2 + y_3 + \cdots + y_k + y_{k+1} \tag{1.108}$$

and

$$\Delta S_k = S_{k+1} - S_k = y_{k+1}. \tag{1.109}$$

The following examples illustrate and extend the use of the above operator relations.

1.6.1. Example A

Let $y_k = 1$; then

$$\Delta \cdot 1 = 1 - 1 = 0. \tag{1.110}$$

Let $y_k = k$; then

$$\Delta \cdot k = (k+1) - k = 1. \tag{1.111}$$

Let $y_k = k^n$, where n is a positive integer; then

$$\begin{aligned} \Delta k^n &= (k+1)^n - k^n \\ &= nk^{n-1} + \binom{n}{2} k^{n-2} + \cdots + \binom{n}{n-1} k + 1. \end{aligned} \tag{1.112}$$

Let $y_k = (-1)^k$; then

$$\Delta(-1)^k = (-1)^{k+1} - (-1)^k = 2(-1)^{k+1}. \tag{1.113}$$

Let $y_k = k(-1)^k$; then

$$\Delta k(-1)^k = (k+1)(-1)^{k+1} - k(-1)^k = -(2k+1)(-1)^k. \tag{1.114}$$

Let $y_k = a^k$; then

$$\Delta a^k = a^{k+1} - a^k = (a-1)a^k \tag{1.115a}$$

and

$$\Delta^n a^k = (a-1)^n a^k. \tag{1.115b}$$

Note that for $a = 2$, we have $\Delta^n 2^k = 2^k$.

Let $y_k = \cos(ak)$; then

$$\begin{aligned} \Delta \cos(ak) &= \cos(ak + a) - \cos(ak) \\ &= \cos a \cos(ak) - \sin a \sin(ak) - \cos(ka) \\ &= (\cos a - 1)\cos(ak) - \sin a \sin(ak). \end{aligned} \tag{1.116}$$

1.6.2. Example B

We now prove the following relations:

$$E(x_k y_k) = (Ex_k)(Ey_k), \tag{1.117}$$

$$E(y_k)^n = (Ey_k)^n. \tag{1.118}$$

From the definition of the shift operator, we have

$$E(x_k y_k) = k_{k+1} y_{k+1}. \tag{1.119}$$

However, $x_{k+1} = Ex_k$ and $y_{k+1} = Ey_k$. Substitution of these results into the right-hand side of equation (1.119) gives equation (1.117).

Again, using the definition of the shift operator, we obtain

$$E(y_k)^n = (y_{k+1})^n = (Ey_k)^n, \tag{1.120}$$

which is just equation (1.118).

Let $f_1(k), f_2(k), \ldots, f_n(k)$ be functions of k. We can conclude that

$$E^m[f_1(k)f_2(k)\cdots f_n(k)] = [E^m f_1(k)]\cdots[E^m f_n(k)]. \tag{1.121}$$

1.6.3. Example C

Let P_k be a polynomial of degree n,

$$P_k = a_0 k^n + a_1 k^{n-1} + \cdots + a_n. \tag{1.122}$$

Prove that

$$\Delta^n P_k = a_0 n! \tag{1.123}$$

and

$$\Delta^{n+m} P_k = 0, \qquad m = 1, 2, \ldots. \tag{1.124}$$

We have

$$\begin{aligned} \Delta P_k &= [a_0(k+1)^n + a_1(k+1)^{n-1} + \cdots + a_n] \\ &\quad - (a_0 k^n + a_1 k^{n-1} + \cdots + a_n) \\ &= a_0 n k^{n-1} + \text{terms of lower degree than } (n-1) \end{aligned} \tag{1.125}$$

and

$$\Delta^2 P_k = a_0 n(n-1)k^{n-2} + \text{terms of degree lower than } (n-2). \quad (1.126)$$

Therefore, each application of the difference operator reduces the degree by one and adds one factor to the succession $n(n-1)(n-2)\cdots$. Carrying out this process n times gives

$$\Delta^n P_k = a_0 n(n-1)(n-2)\cdots(1) = a_0 n! \quad (1.127)$$

Since the right-hand side of equation (1.127) is a constant, the application of addition powers of the Δ operator will give zero.

1.6.4. Example D

Show that

$$(k+r)^n - \binom{r}{1}(k+r-1)^n + \binom{r}{2}(k+r-2)^n + \cdots + (-1)^r \binom{r}{r} k^n = \begin{cases} 0, & \text{for } r > n, \\ n!, & \text{for } r = n. \end{cases} \quad (1.128)$$

We proceed by calculating $\Delta^r k^n$, i.e.,

$$\begin{aligned} \Delta^r k^n &= (E-1)^r k^n \\ &= \left[E^r - \binom{r}{1}E^{r-1} + \binom{r}{2}E^{r-2} + \cdots + (-1)^r\binom{r}{r}\right] k^n. \end{aligned} \quad (1.129)$$

Using the results contained in Example 1.6.3 and the definition of the shift operator, we obtain equation (1.128).

1.6.5. Example E

The quantities $\Delta^r(0^n)$ are called the differences of zero. They can be easily calculated from the results of Example 1.6.4 by setting $k = 0$. Doing this gives

$$\Delta^r(0^n) = \begin{cases} 0, & \text{for } r > n, \\ n!, & \text{for } r = n. \end{cases} \quad (1.130)$$

1.6.6. Example F

Let us use Leibnitz's theorem to calculate $\Delta^n(k^2 v_k)$. From equation (1.99), we have, for $x_k = k^2$ and $y_k = v_k$, the result

$$\Delta^n(k^2 v_k) = k^2 \Delta^n v_k + \binom{n}{1}(\Delta k^2)(\Delta^{n-1} v_{k+1}) + \binom{n}{2}(\Delta^2 k^2)(\Delta^{n-2} v_{k+2}), \quad (1.131)$$

where all other terms are zero since $\Delta^m k^2 = 0$ for $m > 2$. Therefore,

$$\Delta^n(k^2 v_k) = k^2 \Delta^n v_k + (2k+1)\Delta^{n-1} v_{k+1} + n(n-1)\Delta^{n-2} v_{k+2}. \quad (1.132)$$

For the particular case of $v_k = a^k$, we obtain

$$\Delta^n(k^2 a^k) = [(a-1)^2 k^2 + na(a-1)(2k+1) + a^2 n(n-1)](a-1)^{n-2} a^k. \quad (1.133)$$

1.7. FACTORIAL POLYNOMIALS

We saw in Example 1.6.1 that taking the first differnce of $y_k = k^n$ leads to the cumbersome expression

$$\Delta k^n = nk^{n-1} + \binom{n}{2} k^{n-2} + \cdots + 1. \quad (1.134)$$

If we wished to determine the difference of a polynomial function of k, then the expressions obtained would be rather complex. We now show that it is possible to define a particular function of k such that its various differences have a simple structure. This function is called the factorial function.

Let n be a positive integer and define $k^{(n)}$ as

$$k^{(n)} \equiv k(k-1)(k-2)\cdots(k-n-1). \quad (1.135)$$

This expression is read "k, n factorial" and can be rewritten in the form

$$k^{(n)} = \frac{k!}{(k-n)!}. \quad (1.136)$$

Note that

$$k^{(k)} = k!, \tag{1.137}$$

and since for $n > k$, $(k - n)!$ is unbounded,

$$k^{(n)} = 0, \qquad n > k. \tag{1.138}$$

Let j be a positive integer; then from the definition of the factorial function, we have that

$$k^{(n)}(k-n)^{(j)} = \frac{k!}{(k-n)!}\frac{(k-n)!}{(k-n-j)!} = \frac{k!}{(k-n-j)!} = k^{(n+j)}. \tag{1.139}$$

If we require equation (1.139) to hold for $n = 0$ and $j > 0$, then

$$k^{(0)}(k-0)^{(j)} = k^{(0+j)} = k^{(j)}, \tag{1.140}$$

and we conclude that $k^{(0)} = 1$. This result is consistent with equation (1.136).

If we now require equation (1.139) to also hold for negative factorial exponents, then for the special case $n = -j > 0$, we have

$$k^{(-n)}(k+n)^{(n)} = k^{(-n+n)} = 1, \tag{1.141}$$

or

$$k^{(-n)} = \frac{1}{(k+n)^{(n)}} = \frac{1}{(k+n)(k+n-1)\cdots(k+1)}. \tag{1.142}$$

Note that, for n negative, equation (1.142) can be written as

$$k^{(n)} = \frac{1}{(k+1)(k+2)\cdots(k-n-1)(k-n)}. \tag{1.143}$$

We now show that

$$\Delta k^{(n)} = nk^{(n-1)}, \tag{1.144}$$

for all integers k. It should be clear from this result that the following relations hold:

$$\Delta^2 k^{(n)} = \Delta(\Delta k^{(n)}) = \Delta n k^{(n-1)} = n\,\Delta k^{(n-1)} = n(n-1)k^{(n-2)}, \tag{1.145}$$

$$\Delta^3 k^{(n)} = \Delta(\Delta^2 k^{(n)}) = n(n-1)(n-2)k^{(n-3)}, \tag{1.146}$$

$$\vdots$$

$$\Delta^n k^{(n)} = n!. \tag{1.147}$$

To prove equation (1.144), there are three cases to consider, namely, $n > 0$, $n < 0$, and $n = 0$. For the last case, we have

$$\Delta k^{(0)} = \Delta 1 = 0. \tag{1.148}$$

For $n > 0$,

$$\begin{aligned} \Delta k^{(n)} &= (k+1)^{(n)} - k^{(n)} \\ &= (k+1)k(k-1)\cdots(k-n+2) - k(k-1)\cdots \\ &\qquad (k-n+2)(k-n+1) \\ &= [(k+1)-(k-n+1)]k(k-1)\cdots(k-n+2) \\ &= nk^{(n-1)}. \end{aligned} \tag{1.149}$$

For $n < 0$,

$$\begin{aligned} \Delta k^{(n)} &= \frac{1}{(k+2)(k+3)\cdots(k-n)(k-n+1)} \\ &\quad - \frac{1}{(k+1)(k+2)\cdots(k-n-1)(k-n)} \\ &= \frac{(k+1)-(k-n+1)}{(k+1)(k+2)\cdots(k-n)(k-n+1)} \\ &= nk^{(n-1)}. \end{aligned} \tag{1.150}$$

Therefore, we conclude that the expression of equation (1.144) is correct for all integer k.

It should be mentioned that factorial functions can also be defined for noninteger values of k by making use of the gamma function. Let x be a continuous variable; then

$$x^{(n)} = \frac{\Gamma(x+1)}{\Gamma(x+1-n)} \tag{1.151}$$

is the proper generalization of the relation given in equation (1.136). However, in this book, there will be no need to use this generalization.

We define a factorial polynomial of degree n to be an expression of the form

$$\phi_k = a_0 k^{(n)} + a_1 k^{(n-1)} + \cdots + a_{n-1} k^{(1)} + a_n, \tag{1.152}$$

where $a_0, a_1, \ldots, a_n$ are constants, $a_0 \neq 0$, and n is a positive integer.

We now show that if $\phi_k = 0$, for all k, then

$$a_0 = a_1 = \cdots = a_n = 0. \tag{1.153}$$

To proceed, we have

$$a_0 k^{(n)} + a_1 k^{(n-1)} + \cdots + a_n = 0, \tag{1.154}$$

for all k. For $k = 0$, we have

$$a_n = 0. \tag{1.155}$$

If we now take the difference of both sides of equation (1.154),

$$n a_0 k^{(n-1)} + (n-1) a_1 k^{(n-2)} + \cdots + a_{n-1} = 0, \tag{1.156}$$

and set $k = 0$, we obtain

$$a_{n-1} = 0. \tag{1.157}$$

After repeatedly differencing and evaluating the resulting expressions at $k = 0$, we conclude that each coefficient must be zero.

A factorial polynomial is unique. This means that if ϕ_k is given by equation (1.152) and also by some other expression, such as

$$\phi_k = b_{-m} k^{(n+m)} + \cdots + b_{-1} k^{(n+1)} + b_0 k^{(n)} + b_1 k^{(n-1)} + \cdots + b_n, \tag{1.158}$$

then

$$b_{-k} = \cdots = b_{-1} = 0 \tag{1.159}$$

and

$$b_i = a_i, \qquad i = 0, 1, \ldots, n. \tag{1.160}$$

To prove this, subtract equation (1.152) from equation (1.158) to obtain

$$b_{-m}k^{(n+m)} + \cdots + b_{-1}k^{(n+1)} + (b_0 - a_0)k^{(n)} + \cdots + (b_n - a_n) = 0. \qquad (1.161)$$

Since this is to be true for all k, we conclude from the results given by equations (1.153) and (1.154) that the relations of equations (1.159) and (1.160) must hold. Thus, a factorial polynomial has a unique set of coefficients.

Let P_k be a function of k. We define for positive n

$$P_k{}^{(n)} \equiv P_k P_{k-1} P_{k-2} \cdots P_{k-n+1} = \prod_{i=0}^{n-1} P_{k-i}, \qquad (1.162)$$

and, for negative n,

$$P_k{}^{(n)} = (P_{k+1}P_{k+2} \cdots P_{k-n})^{-1} = 1 \bigg/ \prod_{i=1}^{-n} P_{k+i}; \qquad (1.163)$$

$P_k{}^{(0)}$ is defined to be one.

Since the factorial functions have such nice properties under differencing, it is often useful to convert an ordinary polynomial to a factorial polynomial. That is, given

$$P_k = a_0k^n + a_1k^{n-1} + \cdots + a_{n-1}k + a_n, \qquad (1.164)$$

we wish to determine coefficients $c_0, c_1, \ldots, c_n$, such that

$$P_k = c_0k^{(n)} + c_1k^{(n-1)} + \cdots + c_{n-1}k^{(1)} + c_n. \qquad (1.165)$$

Now, from the definition of $k^{(n)}$, given by equation (1.135), we find on setting $n = 0, 1, 2, \ldots$, and expanding

$$\begin{aligned}
k^{(0)} &= 1,\\
k^{(1)} &= k,\\
k^{(2)} &= k^2 - k,\\
k^{(3)} &= k^3 - 3k^2 + 2k,\\
k^{(4)} &= k^4 - 6k^3 + 11k^2 - 6k,\\
k^{(5)} &= k^5 - 10k^4 + 35k^3 - 50k^2 + 24k,
\end{aligned} \qquad (1.166)$$

etc.

These relations can be inverted to give the various powers of k in terms of the factorial functions

$$\begin{aligned}
1 &= k^{(0)}, \\
k &= k^{(1)}, \\
k^2 &= k^{(2)} + k^{(1)}, \\
k^3 &= k^{(3)} + 3k^{(2)} + k^{(1)}, \\
k^4 &= k^{(4)} + 7k^{(3)} + 6k^{(2)} + k^{(1)}, \\
k^5 &= k^{(5)} + 15k^{(4)} + 25k^{(3)} + 10k^{(2)} + k^{(1)}, \\
&\text{etc.}
\end{aligned} \tag{1.167}$$

The expressions given in equations (1.166) and (1.167) can be written in the following forms, respectively:

$$k^{(n)} = \sum_{i=1}^{n} s_i{}^n k^i \tag{1.168}$$

and

$$k^n = \sum_{i=1}^{n} S_i{}^n k^{(i)}. \tag{1.169}$$

The coefficients $s_i{}^n$ are called Stirling numbers of the first kind, while the coefficients $S_i{}^n$ are called Stirling numbers of the second kind. These numbers satisfy the recursion relations

$$s_i{}^{n+1} = s_{i-1}{}^n - n s_i{}^n, \tag{1.170}$$

where for $n > 0$,

$$s_n{}^n = 1, \qquad s_i{}^n = 0, \qquad \text{for } i \leq 0 \text{ and } i \geq n+1, \tag{1.171}$$

and

$$S_i{}^{n+1} = S_{i-1}{}^n + i S_i{}^n, \tag{1.172}$$

where for $n > 0$,

$$S_n{}^n = 1, \qquad S_i{}^n = 0, \qquad \text{for } i \leq 0 \text{ and } i \geq n+1. \tag{1.173}$$

In Example 1. 7.4, we derive equations (1.170) and (1.172).

1.7.1. Example A

Calculate the first difference of $(a + bk)^{(n)}$, where a and b are constants. By definition

$$(a + bk)^{(n)} = (a + bk)[a + b(k - 1)] \cdots [a + b(k - n + 1)]. \tag{1.174}$$

Therefore,

$$\begin{aligned} \Delta(a + bk)^{(n)} &= [a + b(k + 1)](a + bk) \cdots [a + b(k - n + 2)] \\ &\quad - (a + bk) \cdots [a + b(k - n + 2)][a \\ &\quad + b(k - n + 1)] \\ &= \{[a + b(k + 1)] - [a + b(k - n + 1)]\} \\ &\quad \times \{(a + bk) \cdots [a + b(k - n + 2)]\} \\ &= bn(a + bk)^{(n-1)}. \end{aligned} \tag{1.175}$$

Likewise, we have

$$\begin{aligned} \Delta^2(a + bk)^{(n)} = \Delta[\Delta(a + bk)^{(n)}] &= bn\Delta(a + bk)^{(n-1)} \\ &= bn[b(n - 1)](a + bk)^{(n-2)} \\ &= b^2 n(n - 1)(a + bk)^{(n-2)} \end{aligned} \tag{1.176}$$

and, in general

$$\begin{gathered} \Delta^m(a + bk)^{(n)} = b^m n(n - 1) \cdots (n - m + 1)(a + bk)^{(n-m)}, \\ 0 < m < n, \end{gathered} \tag{1.177}$$

$$\Delta^n(a + bk)^{(n)} = b^n n!, \tag{1.178}$$

$$\Delta^p(a + bk)^{(n)} = 0, \qquad \text{for } p > n. \tag{1.179}$$

Equation (1.174) should be used with care as the following examples show. Let us evaluate $P_k = k^{(3)}$ at $k = 6$. From equation (1.174), we have

$$P_k = k^{(3)} = k(k - 1)(k - 2), \tag{1.180}$$

and

$$P_6 = 6(6 - 1)(6 - 2) = 120. \tag{1.181}$$

Now consider

$$Q_k = (1 + 4k)^{(3)}, \tag{1.182}$$

which can be correctly written, using equation (1.174), as

$$\begin{aligned} Q_k &= (1 + 4k)[1 + 4(k - 1)][1 + 4(k - 2)] \\ &= (1 + 4k)(-3 + 4k)(-7 + 4k). \end{aligned} \tag{1.183}$$

Therefore,

$$Q_5 = (1 + 20)(-3 + 20)(-7 + 20) = 4641. \tag{1.184}$$

However, if we had written

$$Q_5 = (1 + 4 \cdot 5)^{(3)} = (21)^{(3)} = (21)(20)(19) = 7980, \tag{1.185}$$

we would obtain the indicated wrong value, 7980, rather than the correct value, 4641. So, be careful!

1.7.2. Example B

Prove the following (Newton's theorem): If f_k is a polynomial of the nth degree, then it can be written in the form

$$f_k = f_0 + \frac{\Delta f_0}{1!} k^{(1)} + \frac{\Delta^2 f_0}{2!} k^{(2)} + \cdots + \frac{\Delta^n f_0}{n!} k^{(n)}. \tag{1.186}$$

Assume that f_k has the representation

$$f_k = a_0 + a_1 k^{(1)} + a_2 k^{(2)} + \cdots + a_n k^{(n)}, \tag{1.187}$$

where $a_0, a_1, \ldots, a_n$ are constants. Differencing f_k n times gives

$$\begin{aligned} \Delta f_k &= a_1 + 2a_2 k^{(1)} + 3a_3 k^{(2)} + \cdots + na_n k^{(n-1)}, \\ \Delta^2 f_k &= 2 \cdot 1 \cdot a_2 + 3 \cdot 2 \cdot a_3 k^{(1)} + \cdots + n(n-1)a_n k^{(n-2)}, \\ &\quad \vdots \\ \Delta^n f_k &= a_n n(n-1) \cdots (1). \end{aligned} \tag{1.188}$$

Setting $k = 0$ in the original function and its differences allows us to conclude that

$$a_m = \frac{\Delta^m f_0}{m!}, \qquad m = 0, 1, \ldots, n. \tag{1.189}$$

To illustrate the use of this theorem consider the function

$$f_k = k^4. \tag{1.190}$$

Now

$$\begin{aligned} \Delta f_k &= 4k^3 + 6k^2 + 4k + 1, \\ \Delta^2 f_k &= 12k^2 + 24k + 14, \\ \Delta^3 f_k &= 24k + 36, \\ \Delta^4 f_k &= 24, \end{aligned} \tag{1.191}$$

and

$$f_0 = 0, \qquad \Delta f_0 = 1, \qquad \Delta^2 f_0 = 14, \qquad \Delta^3 f_0 = 36, \qquad \Delta^4 f_0 = 24. \tag{1.192}$$

Therefore, from equation (1.189),

$$a_0 = 0, \qquad a_1 = 1, \qquad a_2 = 7, \qquad a_3 = 6, \qquad a_4 = 1, \tag{1.193}$$

and $f_k = k^4$ has the factorial polynomial representation

$$f_k = k^4 = k^{(1)} + 7k^{(2)} + 6k^{(3)} + k^{(4)}. \tag{1.194}$$

For a second example consider

$$f_k = 3k^3 + k - 1. \tag{1.195}$$

We have

$$\Delta f_k = 9k^2 + 9k + 4, \qquad \Delta^2 f_k = 18k + 18, \qquad \Delta^3 f_k = 18, \tag{1.196}$$

and, from equation (1.189),

$$a_0 = -1, \qquad a_1 = 4, \qquad a_2 = 9, \qquad a_3 = 3. \tag{1.197}$$

Therefore,

$$f_k = 3k^3 + k - 1 = 3k^{(3)} + 9k^{(2)} + 4k^{(1)} - 1. \tag{1.198}$$

1.7.3. Example C

Use the expansions in equation (1.167) to obtain the factorial polynomial representation of the function

$$P_k = k^5 - 4k^4 + 2k^3 - k^2 + k + 10. \tag{1.199}$$

Substituting directly from equations (1.167) gives

$$\begin{aligned} P_k = {} & (k^{(5)} + 15k^{(4)} + 25k^{(3)} + 10k^{(2)} + k^{(1)}) \\ & -4(k^{(4)} + 7k^{(3)} + 6k^{(2)} + k^{(1)}) \\ & +2(k^{(3)} + 3k^{(2)} + k^{(1)}) - (k^{(2)} + k^{(1)}) + k^{(1)} + 10 \end{aligned} \tag{1.200}$$

which on adding similar terms gives

$$P_k = k^{(5)} + 11k^{(4)} - k^{(3)} - 9k^{(2)} - k^{(1)} + 10. \tag{1.201}$$

1.7.4. Example D

We now derive the recursion relation for Stirling numbers of the first kind.

Using the fact that $s_n{}^n = 1$, $s_i{}^n = 0$, for $i \leq 0$, $i \geq n + 1$, where $n > 0$, we can rewrite equation (1.168) as follows:

$$k^{(n)} = \sum_{i=-\infty}^{\infty} s_i{}^n k^i. \tag{1.202}$$

Therefore,

$$k^{(n+1)} = \sum_{i=-\infty}^{\infty} s_i{}^{n+1} k_i. \tag{1.203}$$

From the definition of the factorial polynomials, we have

$$k^{(n+1)} = (k - n)k^{(n)}. \tag{1.204}$$

Substituting equations (1.202) and (1.203) into equation (1.204) gives

$$\sum_{i=-\infty}^{\infty} s_i^{n+1}k^i = (k-n)\sum_{i=-\infty}^{\infty} s_i^n k^i$$

$$= \sum_{i=-\infty}^{\infty} s_i^n k^{i+1} - \sum_{i=-\infty}^{\infty} ns_i^n k^i \qquad (1.205)$$

$$= \sum_{i=-\infty}^{\infty} s_{i-1}^n k^i - \sum_{i=-\infty}^{\infty} ns_i^n k^i.$$

If we equate the coefficients of k^i, we obtain

$$s_i^{n+1} = s_{i-1}^n - ns_k^n, \qquad (1.206)$$

which is just equation (1.170).

The defining equation for Stirling numbers of the second kind can be written as

$$k^n = \sum_{i=-\infty}^{\infty} S_i^n k^{(i)}, \qquad (1.207)$$

if the conditions given by equation (1.173) are used. Now

$$k^{n+1} = \sum_{k=-\infty}^{\infty} S_i^{n+1} k^{(i)}, \qquad (1.208)$$

and

$$k^{n+1} = kk^n. \qquad (1.209)$$

Substitution of equations (1.207) and (1.208) into equation (1.209) gives

$$\sum_{i=-\infty}^{\infty} S_i^{n+1} k^{(i)} = k \sum_{i=-\infty}^{\infty} S_i^n k^{(i)} = \sum_{i=-\infty}^{\infty} S_i^n kk^{(i)}. \qquad (1.210)$$

However,

$$kk^{(i)} = kk(k-1)\cdots(k-i+1)$$

$$= (k-i+i)k(k-1)\cdots(k-i+1) \qquad (1.211)$$

$$= k^{(i+1)} + ik^{(i)}.$$

Using this last result in equation (1.210), we obtain

$$\sum_{i=-\infty}^{\infty} S_i^{n+1} k^{(i)} = \sum_{i=-\infty}^{\infty} S_i^n (k^{(i+1)} + ik^{(i)})$$

$$= \sum_{i=-\infty}^{\infty} S_i^n k^{(i+1)} + \sum_{i=-\infty}^{\infty} iS_i^n k^{(i)} \tag{1.212}$$

$$= \sum_{i=-\infty}^{\infty} S_{i-1}^n k^{(i)} + \sum_{i=-\infty}^{\infty} iS_i^n k^{(i)},$$

and thus can conclude that

$$S_i^{n+1} = S_{i-1}^n + iS_i^n, \tag{1.213}$$

which is just equation (1.172).

1.8. OPERATOR Δ^{-1} AND THE SUM CALCULUS

We define $\Delta^{-1} y_k$ to be such that

$$\Delta(\Delta^{-1} y_k) = y_k. \tag{1.214}$$

Let $z_k = \Delta^{-1} y_k$; then from equation (1.214)

$$\Delta z_k = z_{k+1} - z_k = y_k, \tag{1.215}$$

and we have

$$\begin{aligned} z_{k+1} - z_k &= y_k, \\ z_k - z_{k-1} &= y_{k-1}, \\ z_{k-1} - z_{k-2} &= y_{k-2}, \\ &\vdots \\ z_2 - z_1 &= y_1. \end{aligned} \tag{1.216}$$

Now adding the left-hand sides and the right-hand sides gives

$$z_{k+1} - z_1 = y_1 + y_2 + \cdots + y_{k-1} + y_k \tag{1.217}$$

or

$$z_{k+1} = z_1 + \sum_{r=1}^{k} y_r, \tag{1.218}$$

and

$$z_k = z_1 + \sum_{r=1}^{k-1} y_r. \tag{1.219}$$

Therefore, substituting $z_k = \Delta^{-1} y_k$, we obtain

$$\Delta^{-1} y_k = \sum_{r=1}^{k-1} y_r + \text{constant}, \tag{1.220}$$

since z_1 is an arbitrary constant.

In summary, $\Delta^{-1} y_k$ is a function whose difference is y_k. In fact, for any positive integer n, we define $\Delta^{-n} y_k$ as a function whose nth difference is y_k.

Now define

$$\Delta^{-n} y_k = \Delta^{-1}(\Delta^{-n+1} y_k). \tag{1.221}$$

Therefore,

$$\begin{aligned} \Delta^{-2} y_k &= \Delta^{-1}(\Delta^{-1} y_k) = \Delta^{-1}\left(\sum_{r=1}^{k-1} y_r + c_1\right) \\ &= \sum_{m=1}^{k-1} \sum_{r=1}^{m-1} y_r + \Delta^{-1} c_1 + c_2, \end{aligned} \tag{1.222}$$

where c_1 and c_2 are aribtrary constants. Now $\Delta^{-1} c_1$ is easy to determine since $\Delta z_k = 1$ implies $z_k = k +$ constant. Consequently,

$$\Delta^{-2} y_k = \sum_{m=1}^{k-1} \sum_{r=1}^{m-1} y_r + c_1 k + c_2. \tag{1.223}$$

In a similar fashion, we have

$$\begin{aligned} \Delta^{-3} y_k &= \Delta^{-1}(\Delta^{-2} y_k) \\ &= \sum_{l=1}^{k-1} \sum_{m=1}^{l-1} \sum_{r=1}^{m-1} y_r + c_1 k^2 + c_2 k + c_3, \end{aligned} \tag{1.224}$$

where c_1, c_2, and c_3 are arbitrary constants. Using an obvious notation, we can generalize the above results to give

$$\Delta^{-n} y_k = (\Sigma)^n y_k + c_1 k^{n-1} + c_2 k^{n-2} + \cdots + c_n, \tag{1.225}$$

where the n constants c_i are arbitrary.

We now show that the operators Δ and Δ^{-1} do not commute, i.e., $\Delta\Delta^{-1} \neq \Delta^{-1}\Delta$. Using the definition of Δ and the above interpretation of Δ^{-1}, we have

$$\begin{aligned} \Delta^{-1}\Delta y_k &= \Delta^{-1}(y_{k+1} - y_k) = c + \sum_{r=1}^{k-1} (y_{k+1} - y_k) \\ &= c + y_k - y_1 = c_1 + y_k, \end{aligned} \tag{1.226}$$

where c, y_1, and c_1 are constants. Replacing y_k on the right-hand side by the definition of equation (1.214) gives

$$\Delta^{-1}\Delta y_k = \Delta\Delta^{-1} y_k + c_1. \tag{1.227}$$

Thus, we conclude that $\Delta^{-1}\Delta y_k$ differs from $\Delta\Delta^{-1} y_k$ by an arbitrary constant. This result is easily seen by considering that for $\Delta^{-1}\Delta$ the summation is at the end of the joint operation and introduces an arbitrary constant. On the other hand, for $\Delta\Delta^{-1}$ the difference operation is last and thus destroys the arbitrary constant introduced by the summation operation.

We now prove a formula for summation by parts. From equation (1.98), we have

$$\Delta(x_k y_k) = x_{k+1}\Delta y_k + y_k \Delta x_k, \tag{1.228}$$

or

$$y_k \Delta x_k = \Delta(x_k y_k) - x_{k+1}\Delta y_k. \tag{1.229}$$

Applying the operator Δ^{-1} to both sides gives

$$\Delta^{-1}(y_k \Delta x_k) = \Delta^{-1}\Delta(x_k y_k) - \Delta^{-1}(x_{k+1}\Delta y_k), \tag{1.230}$$

and

$$\sum_{r=1}^{k-1} y_r \Delta x_r = x_k y_k - \sum_{r=1}^{k-1} x_{r+1}\Delta y_r + \text{constant}. \tag{1.231}$$

The last relation is the required formula for summation by parts.

The following result is known as the fundamental theorem of the sum calculus.

Theorem 1.4: If $\Delta F_k = f_k$ and a, $b \geq a$ are integers,

$$\sum_{k=a}^{b} f_k = F(b+1) - F(a). \tag{1.232}$$

Proof: This theorem follows directly from the expression given by equation (1.219).

We now derive the so-called Abel transformation

$$\sum_{k=1}^{n} x_k y_k = x_{n+1} \sum_{k=1}^{n} y_k - \sum_{k=1}^{n} \left(\Delta x_k \sum_{r=1}^{k} y_r \right). \tag{1.233}$$

Using the relation for summation by parts given by equation (1.231) and the fundamental theorem of the sum calculus, we obtain the following:

$$\sum_{k=1}^{n} f_k \Delta g_k = f_{n+1} g_{n+1} - f_1 g_1 - \sum_{k=1}^{n} g_{k+1} \Delta f_k, \tag{1.234}$$

where f_k and g_k are functions of k. Now define

$$f_k = x_k, \quad \Delta g_k = y_k. \tag{1.235}$$

Consequently,

$$\sum_{k=1}^{n-1} \Delta g_k = g_n - g_1 = \sum_{k=1}^{n-1} y_k, \tag{1.236}$$

and

$$g_n = g_1 + \sum_{k=1}^{n-1} y_k. \tag{1.237}$$

Substitution of these results into equation (1.234) gives

$$\begin{aligned} \sum_{k=1}^{n} x_k y_k &= x_{k+1} \left(g_1 + \sum_{k=1}^{n} y_k \right) - x_1 g_1 \\ &\quad - \sum_{k=1}^{n} \left[\Delta x_k \left(g_1 + \sum_{r=1}^{k} y_r \right) \right] \\ &= x_{n+1} \sum_{k=1}^{n} y_k - \sum_{k=1}^{n} \left(\Delta x_k \sum_{r=1}^{k} y_r \right) \\ &\quad + x_{n+1} g_1 - x_1 g_1 - \sum_{k=1}^{n} g_1 \Delta x_k. \end{aligned} \tag{1.238}$$

Now

$$x_{n+1}g_1 - x_1 g_1 - \sum_{k=1}^{n} g_1 \Delta x_k = x_{n+1}g_1 - x_1 g_1 - g_1 \sum_{k=1}^{n} (x_{k+1} - x_k) = 0. \tag{1.239}$$

Therefore, we obtain the result given by equation (1.233).

We now turn to a consideration of the operator $(1-\lambda\Delta)^{-1}$, where λ is a constant. We define $(1-\lambda\Delta)^{-1}y_k$ as follows:

$$(1-\lambda\Delta)(1-\lambda\Delta)^{-1}y_k = y_k. \tag{1.240}$$

For a given y_k, $\Delta^n y_k$ is some function of k. With this in mind, consider the series

$$y_k + \lambda\Delta y_k + \lambda^2\Delta^2 y_k + \cdots, \tag{1.241}$$

and assume that it either converges absolutely or has only a finite number of terms. (The latter case will occur whenever y_k is a polynomial function.) Note that this series corresponds to the formal expansion of the operator $(1-\lambda\Delta)^{-1}$ acting on y_k; that is,

$$\begin{aligned}(1-\lambda\Delta)^{-1}y_k &= \frac{1}{1-\lambda\Delta}y_k \\ &= (1+\lambda\Delta+\lambda^2\Delta^2+\cdots)y_k \\ &= y_k + \lambda\Delta y_k + \lambda^2\Delta^2 y_k + \cdots.\end{aligned} \tag{1.242}$$

Therefore, if

$$(1-\lambda\Delta)z_k = y_k, \tag{1.243}$$

then a value of z_k can be determined by expanding $(1-\lambda\Delta)^{-1}$ by the binomial theorem and letting the series operate on y_k,

$$\begin{aligned}z_k &= (1-\lambda\Delta)^{-1}y_k \\ &= y_k + \lambda\Delta y_k + \lambda^2\Delta^2 y_k + \cdots.\end{aligned} \tag{1.244}$$

As we will see in Chapter 4, this procedure does not give the general solution for z_k; however, it does provide a unique expression for z_k given y_k. We take this to be the unique meaning to attach to the expression $(1 - \lambda\Delta)^{-1}y_k$. Finally, it is easy to show that the operators $(1 - \lambda\Delta)$ and $(1 - \lambda\Delta)^{-1}$ commute; that is,

$$(1-\lambda\Delta)(1-\lambda\Delta)^{-1}y_k = (1-\lambda\Delta)^{-1}(1-\lambda\Delta)y_k. \tag{1.245}$$

The following two theorems permit the use of operator methods to sum series.

Theorem 1.5: Consider the absolutely convergent series

$$S(x) = a_0 + a_1x + a_2x^2 + \cdots + a_kx^k + \cdots, \tag{1.246}$$

where a_k is a given function of k. Then $S(x)$ can be expressed in the form

$$S(x) = \frac{a_0}{1-x} + \frac{x\Delta a_0}{(1-x)^2} + \frac{x^2\Delta^2 a_0}{(1-x)^3} + \cdots. \tag{1.247}$$

This result is known as Montmort's theorem on infinite summation. Note that if a_k is a polynomial in k of degree n, then $\Delta^m a_0$ will be zero for all $m > n$ and thus a finite number of terms for the series $S(x)$ will occur.

Proof: We have

$$\begin{aligned} S(x) &= a_0 + a_1x + a_2x^2 + \cdots + a_kx^k + \cdots \\ &= (1 + xE + x^2E^2 + \cdots + x^kE^k + \cdots)a_0 \\ &= (1 - xE)^{-1}a_0 = [1 - x(1+\Delta)]^{-1}a_0 \\ &= \frac{1}{1-x}\left(1 - \frac{x}{1-x}\Delta\right)^{-1} a_0 \\ &= \frac{1}{1-x}\left(1 + \frac{x\Delta}{1-x} + \frac{x^2\Delta^2}{(1-x)^2} + \cdots\right) a_0 \\ &= \frac{a_0}{1-x} + \frac{x\Delta a_0}{(1-x)^2} + \frac{x^2\Delta^2 a_0}{(1-x)^3} + \cdots. \end{aligned} \tag{1.248}$$

Theorem 1.6: For $\lambda \neq 1$,

$$\sum \lambda^k P_k = \frac{\lambda^k}{\lambda - 1}\left(1 - \frac{\lambda\Delta}{\lambda-1} + \frac{\lambda^2\Delta^2}{(\Delta-1)^2} - \cdots\right) P_k. \tag{1.249}$$

Proof: Let F_k be a function of k. Therefore

$$\begin{aligned}\Delta\lambda^k F_k &= \lambda^{k+1}F_{k+1} - \lambda^k F_k \\ &= \lambda^{k+1}EF_k - \lambda^k F_k \\ &= \lambda^k(\lambda E - 1)F_k.\end{aligned} \tag{1.250}$$

Now, set $(\lambda E - 1)F_k = P_k$; consequently

$$F_k = (\lambda E - 1)^{-1}P_k. \tag{1.251}$$

Therefore, from equation (1.250)

$$\Delta\lambda^k F_k = \lambda^k P_k, \tag{1.252}$$

and

$$\begin{aligned}\Delta^{-1}\lambda^k P_k &= \lambda^k F_k = \lambda^k(\lambda E - 1)^{-1}P_k \\ &= \lambda^k \frac{1}{\lambda(1+\Delta) - 1} P_k \\ &= \frac{\lambda^k}{\lambda - 1} \frac{1}{1 + \lambda\Delta/(\lambda - 1)} P_k \\ &= \frac{\lambda^k}{\lambda - 1}\left(1 - \frac{\lambda\Delta}{\lambda - 1} + \frac{\lambda^2\Delta^2}{(\lambda - 1)^2} - \cdots\right) P_k.\end{aligned} \tag{1.253}$$

1.8.1. Example A

We give in Tables 1.1 and 1.2 a short listing of the antidifferences and definite sums of selected functions. In each case, the particular item is calculated using the definition of $\Delta^{-1}y_k$ and the fundamental theorem of the sum calculus. For example,

$$\Delta^{-1}1 = \sum_{r=1}^{k-1} 1 = k + \text{constant}. \tag{1.254}$$

The arbitrary constants are not indicated in Table 1.1. Likewise, from the fundamental theorem of calcus, we have

Table 1.1
Selected Functions and Their Antidifferences

NUMBER	y_k	$\Delta^{-1}y_k$
1	1	k
2	a^k	$\frac{a^k}{a-1}$, $a \neq 1$
3	$(-1)^k$	$\frac{1}{2}(-1)^{k+1}$
4	ka^k	$\frac{a^k}{a-1}\left(k - \frac{a}{a-1}\right)$, $a \neq 1$
5	$(ak+b)^{(n)}$	$\frac{(ak+b)^{(n+1)}}{a(n+1)}$
6	$\sin(ak+b)$	$-\frac{\cos(ak+b-a/2)}{2\sin(a/2)}$
7	$\cos(ak+b)$	$\frac{\sin(ak+b-a/2)}{2\sin(a/2)}$

$$\sum_{k=0}^{n} 1 = \Delta^{-1}1\Big|_0^{n+1} = n+1. \tag{1.255}$$

For a second example, consider $y_k = a^k$. We wish to determine $\Delta^{-1}a^k$. Now

$$\Delta a^k = (a-1)a^k. \tag{1.256}$$

Applying Δ^{-1} to both sides and dividing by $a - 1$ gives

$$\Delta^{-1}a_k = \frac{a^k}{a-1}, \tag{1.257}$$

provided $a \neq 1$. Also,

$$\sum_{k=0}^{n} a^k = \Delta^{-1}a^k\Big|_0^{n+1} = \frac{a^{n+1}-1}{a-1}, \qquad a \neq 1, \tag{1.258}$$

which is item number 7 in Tables 1.2.

By proceeding in this manner, all the items in both tables can be determined.

Table 1.2
Definite Sums of Selected Functions

NUMBER	SUMMATION	DEFINITE SUM
1	$\sum_{k=0}^{n} 1$	$n+1$
2	$\sum_{k=0}^{n} k$	$\frac{n(n+1)}{2}$
3	$\sum_{k=0}^{n} k^2$	$\frac{n(n+1)(2n+1)}{6}$
4	$\sum_{k=0}^{n} k^3$	$\frac{n^2(n+1)^2}{4}$
5	$\sum_{k=0}^{n} k^{(m)},\ m \neq -1$	$\frac{(n+1)^{(m+1)}}{m+1}$
6	$\sum_{k=0}^{n} (ak+b)^{(m)},\ m \neq -1$	$\frac{[a(n+1)+b]^{(m+1)} - b^{(m+1)}}{a(m+1)}$
7	$\sum_{k=0}^{n} a^k,\ a \neq 1$	$\frac{a^{n+1}-1}{a-1}$
8	$\sum_{k=0}^{n} ka^k,\ a \neq 1$	$\frac{(a-1)(n+1)a^{n+1} - a^{n+2} + a}{(a-1)^2}$
9	$\sum_{k=0}^{n} \sin(ak),\ a \neq 2m\pi$	$\frac{\cos\{a[(n+1)/2]\}\cos(an/2)}{\sin(a/2)}$
10	$\sum_{k=0}^{n} \cos(ak),\ a \neq 2m\pi$	$\frac{\sin\{a[(n+1)/2]\}\cos(an/2)}{\sin(a/2)}$

1.8.2. Example B

Calculate the sum $\sum_{k=1}^{n} k^4$. From numbers 5 of Table 1.1, for $a = 1$, $b = 0$, we obtain

$$\Delta^{-1} k^{(n)} = \frac{k^{(n+1)}}{n+1}. \tag{1.259}$$

Now, from equation (1.220), for $y_k = k^4$, we have

$$\Delta^{-1}(n+1)^4 = \sum_{r=1}^{n} r^4 + \text{constant}, \tag{1.260}$$

or

$$\sum_{r=1}^{n} r^4 = \Delta^{-1}(n+1)^4 + A, \tag{1.261}$$

where A is a constant whose value will be determined later. The term $(n+1)^4$ is a polynomial of degree four and can be expressed as a sum of factorial functions:

$$\begin{aligned}(n+1)^4 &= c_0 + c_1 n^{(1)} + c_2 n^{(2)} + c_3 n^{(3)} + c_4 n^{(4)} \\ &= c_0 + c_1 n + c_2 n(n-1) + c_3 n(n-1)(n-2) \\ &\quad + c_4 n(n-1)(n-2)(n-3).\end{aligned} \tag{1.262}$$

Setting $n = 0$ gives $c_0 = 1$; setting $n = 1$ gives $c_1 + c_0 = 16$ and $c_1 = 15$; setting $n = 2$ gives $c_2 = 25$; setting $n = 3$ gives $c_3 = 10$; finally, we obtain $c_4 = 1$. Therefore,

$$(n+1)^4 = n^{(4)} + 10n^{(3)} + 25n^{(2)} + 15n^{(1)} + 1, \tag{1.263}$$

and applying Δ^{-1} to both sides and using equation (1.259) gives

$$\Delta^{-1}(n+1)^4 = \tfrac{1}{5}n^{(5)} + \tfrac{5}{2}n^{(4)} + \tfrac{25}{3}n^{(3)} + \tfrac{15}{2}n^{(2)} + n^{(1)}, \tag{1.264}$$

or

$$\sum_{r=1}^{n} r^4 = \tfrac{1}{5}n^{(5)} + \tfrac{5}{2}n^{(4)} + \tfrac{25}{3}n^{(3)} + \tfrac{15}{2}n^{(2)} + n^{(1)} + A. \tag{1.265}$$

The constant A can be determined by taking $n = 1$ and obtaining $1 = 1 + A$ or $A = 0$. Finally, we have the result

$$\sum_{r=1}^{n} r^4 = \tfrac{1}{5}n^{(5)} + \tfrac{5}{2}n^{(4)} + \tfrac{25}{3}n^{(3)} + \tfrac{15}{2}n^{(2)} + n^{(1)}. \tag{1.266a}$$

We can use equation (1.166) to express the factorial functions as powers of n; doing this gives

$$\sum_{r=1}^{n} r^4 = \frac{n(6n^4 + 15n^3 + 10n^2 - 1)}{30}. \tag{1.266b}$$

1.8.3. Example C

Determine the quantity $\Delta^{-1}(k3^k)$. Let $y_k = k$ and $\Delta x_k = 3^k$; therefore, $\Delta y_k = 1$ and $x_k = \Delta^{-1}3^k = 3^k/2$. Putting these into the formula for summation by parts, equation (1.231), gives

$$\begin{aligned} \Sigma\, k3^k &= \tfrac{1}{2}k3^k - \Sigma\, \tfrac{1}{2}3^{k+1} + c \\ &= \tfrac{1}{2}k3^k - \tfrac{3}{4}\cdot 3^k + c \\ &= (\tfrac{1}{2}k - \tfrac{3}{4})3^k + c \end{aligned} \tag{1.267}$$

where c is a constant.

1.8.4. Example D

Find the sum of the series $1\cdot 2 + 2\cdot 3 + 3\cdot 4 + \cdots + n(n+1)$.

We have $k(k+1) = (k+1)^{(2)}$; therefore, from the fundamental theorem of the summation calculus, we obtain

$$\begin{aligned} \sum_{k=1}^{n} k(k+1 = \sum_{k=1}^{n}(k+1)^{(2)} &= \Delta^{-1}(k+1)^{(2)}\Big|_{1}^{n+1} \\ &= \tfrac{1}{3}(k+1)^{(3)}\Big|_{1}^{n+1} \\ &= \tfrac{1}{3}(n+2)^{(3)} - \tfrac{1}{3}\cdot 2^{(3)} \\ &= \tfrac{1}{3}(n+2)(n+1)n. \end{aligned} \tag{1.268}$$

1.8.5. Example E

Show that

$$\Sigma \sin(ak) = -\frac{\cos(ak - a/2)}{2\sin(a/2)}. \tag{1.269}$$

Taking the difference of $\cos(ak)$,

$$\Delta\cos(ak) = -2\sin(a/2)\sin(ak + a/2), \tag{1.270}$$

and replacing k by $k - \tfrac{1}{2}$ gives

$$\Delta\cos(ak - a/2) = -2\sin(a/2)\sin(ak). \tag{1.271}$$

Now dividing by $-2\sin(a/2)$ and applying the operator Δ^{-1} gives

$$\Delta^{-1}\sin(ak) = -\frac{\cos(ak - a/2)}{2\sin(a/2)}, \tag{1.272}$$

which, up to a constant, is equation (1.269).

1.8.6. Example F

Use Theorem 1.6 to evaluate $\sum_{k=1}^{n} k \cdot 2^k$.

Comparison of this expression with equation (1.249) gives $\lambda = 2$ and $P_{k,} = k$. Since $\Delta P_k = 1$ and $\Delta^m P_k = 0$ for $m > 1$, we have from equation (1.249)

$$\begin{aligned}\Delta^{-1}2^k k = \Sigma\, 2^k k &= \frac{2^k}{2-1}\left(k - \frac{2\cdot 1}{2-1} + 0 + \cdots + 0 + \cdots\right) \\ &= 2^k(k-2).\end{aligned} \tag{1.273}$$

Applying the fundamental theorem of the sum calculus gives

$$\begin{aligned}\sum_{k=1}^{n} k2^k &= \Delta^{-1}(k2^k)\big|_1^{n+1} \\ &= 2^k(k-2)\big|_1^{n+1} \\ &= 2^{n+1}(n-1) + 2.\end{aligned} \tag{1.274}$$

1.8.7. Example G

Consider the series

$$S(x) = x + 2x^2 + 3x^3 + \cdots + kx^k + \cdots, \tag{1.275}$$

where the coefficient of x^k is $a_k = k$. Therefore, $\Delta a_k = 1$, $\Delta^m a_k = 0$ for $m \geq 2$; and $a_0 = 0$, $\Delta a_0 = 1$, $\Delta^m a_0 = 0$ for $m \geq 2$. We can apply equation (1.247) of Theorem 1.5 to obtain a closed expression for $S(x)$. It is

$$\begin{aligned}S(x) &= \frac{0}{1-x} + \frac{x\cdot 1}{(1-x)^2} + \frac{0}{(1-x)^3} + \cdots + 0 + \cdots \\ &= \frac{x}{(1-x)^2}.\end{aligned} \tag{1.276}$$

1.8.8. Example H

We introduce another technique for summing series that have the form

$$S(x) = a_0 + \frac{a_1 x}{1!} + \frac{a_2 x^2}{2!} + \cdots + \frac{a_k x^k}{k!} + \cdots, \tag{1.277}$$

where a_k is a function of k. Using the shift operator, we have

$$a_k = E^k a_0, \tag{1.278}$$

and equation (1.277) can be written

$$\begin{aligned} S(x) &= \left(1 + \frac{xE}{1!} + \frac{x^2E^2}{2!} + \cdots + \frac{x^kE^k}{k!} + \cdots\right) a_0 \\ &= e^{xE} a_0 = e^{x(1+\Delta)} a_0 = e^x e^{x\Delta} a_0 \\ &= e^x \left(a_0 + \frac{x\Delta a_0}{1!} + \frac{x^2 \Delta^2 a_0}{2!} + \cdots\right). \end{aligned} \tag{1.279}$$

If a_k is a polynomial function of k of nth degree, then $\Delta^m a_k = 0$ for $m > n$ and the right-hand side of equation (1.279) has only a finite number of terms.

To illustrate this, let $a_k = k^2 - 1$, so that

$$S(x) = -1 + \frac{3x^2}{2!} + \frac{8x^3}{3!} + \cdots + \frac{(k^2-1)x^k}{k!} + \cdots. \tag{1.280}$$

Now $\Delta a_k = 2k + 1$, $\Delta^2 a_k = 2$, and $\Delta^m a_k = 0$ for $m > 2$. Therefore, $a_0 = -1$, $\Delta a_0 = 1$, $\Delta^2 a_0 = 2$, and all higher differences are zero. Substitution of these results into equation (1.279) gives

$$\begin{aligned} S(x) &= e^x \left(-1 + \frac{x \cdot 1}{1!} + \frac{x^2 \cdot 2}{2!} + 0 + \cdots + 0 + \cdots\right) \\ &= e^x(x^2 + x - 1), \end{aligned} \tag{1.281}$$

which is the required summed series.

2
FIRST-ORDER DIFFERENCE EQUATIONS

2.1. INTRODUCTION

In this chapter, we study in detail a variety of methods for obtaining solutions to first-order difference equations. Section 2.2 presents a general method for constructing the solution to the linear equation with variable coefficients. The next two sections are concerned with particular forms of the linear difference equation. In Section 2.5, we show how continued fractions can be used to obtain solutions in the form of a series. Finally, in Sections 2.7 and 2.8, we investigate the general nonlinear, first-order difference equation by means of geometric and expansion techniques.

2.2. GENERAL LINEAR EQUATION

The general form of the linear difference equation of first order is

$$y_{k+1} - p_k y_k = q_k, \tag{2.1}$$

where p_k and q_k are given functions. If q_k is identically equal to zero, then the homogeneous equation

$$y_{k+1} - p_k y_k = 0 \tag{2.2}$$

is obtained. Otherwise, equation (2.1) is an inhomogeneous equation. The general solution of equation (2.1) consists of the sum of the solution to the homogeneous equation (2.2) and any particular solution of the inhomogeneous equation (2.1).

We will now show that the general solution to equation (2.1) can be found in finite form.

Consider first the homogeneous equation (2.2). Note that if y_1 is given, then

$$
\begin{aligned}
y_2 &= p_1 y_1, \\
y_3 &= p_2 y_2, \\
&\ \ \vdots \\
y_{k-1} &= p_{k-2} y_{k-2}, \\
y_k &= p_{k-1} y_{k-1}.
\end{aligned}
\tag{2.3}
$$

Multiplying left- and right-hand sides together and setting the two equal gives

$$
\begin{aligned}
y_k &= y_1 p_1 p_2 \cdots p_{k-2} p_{k-1} \\
&= y_1 \prod_{i=1}^{k-1} p_i.
\end{aligned}
\tag{2.4}
$$

Since y_1 can take any value that we wish, equation (2.4) is the general solution to equation (2.2).

Now consider the inhomogeneous equation (2.1). Dividing both left- and right-hand sides by $\prod_{i=1}^{k} p_i$ gives

$$
y_{k+1} \Big/ \prod_{i=1}^{k} p_i - y_k \Big/ \prod_{i=1}^{k-1} p_i = q_k \Big/ \prod_{i=1}^{k} p_i, \tag{2.5}
$$

which can be written

$$
\Delta \left(y_k \Big/ \prod_{i=1}^{k-1} p_i \right) = q_k \Big/ \prod_{i=1}^{k} p_i. \tag{2.6}
$$

Therefore, a particular solution of equation (2.1) is

$$
y_k \Big/ \prod_{i=1}^{k-1} p_i = \Delta^{-1} \left(q_k \Big/ \prod_{i=1}^{k} p_i \right), \tag{2.7}
$$

which can be expressed as

$$
y_k = \left(\prod_{i=1}^{k-1} p_i \right) \sum_{i=1}^{k-1} \left(q_i \Big/ \prod_{r=1}^{i} p_r \right). \tag{2.8}
$$

The general solution of equation (2.1) is the sum of the homogeneous solution and the particular solution,

$$y_k = A \prod_{i=1}^{k-1} p_i + \left(\prod_{i=1}^{k-1} p_i \right) \sum_{i=1}^{k-1} \left(q_i \Big/ \prod_{r=1}^{i} p_r \right), \tag{2.9}$$

where A is an arbitrary constant.

2.2.1. Example A

The first-order, homogeneous difference equation with constant coefficients has the form

$$y_{k+1} - \beta y_k = 0, \qquad \beta = \text{constant}. \tag{2.10}$$

This corresponds to $p_k = \beta$. Therefore, from equation (2.4), the solution is

$$y_k = A \prod_{i=1}^{k-1} \beta = C\beta^k, \tag{2.11}$$

where $C = A/\beta$ is an arbitrary constant.

2.2.2. Example B

Now consider the inhomogeneous equation

$$y_{k+1} - \beta y_k = \alpha, \tag{2.12}$$

where α and β are constants. For this case, we have $p_k = \beta$ and $g_k = \alpha$. Therefore,

$$\sum_{i=1}^{k-1} \left(q_i \Big/ \prod_{r=1}^{i} p_r \right) = \sum_{i=1}^{k-1} \frac{\alpha}{\beta^i} = \alpha \sum_{i=1}^{k-1} \beta^{-i}. \tag{2.13}$$

Using the fact that

$$\sum_{i=1}^{k-1} r^i = \frac{r - r^k}{1 - r} \tag{2.14}$$

gives

$$\sum_{i=1}^{k-1} \beta^{-i} = \frac{1 - \beta^{-k+1}}{\beta - 1}. \tag{2.15}$$

Therefore, for $\beta \neq 1$, the general solution is

$$y_k = C\beta^k - \frac{\alpha}{\beta - 1}, \tag{2.16}$$

where C is an arbitrary constant.

When $\beta = 1$, we have

$$y_{k+1} - y_k = \alpha, \tag{2.17}$$

with $p_k = 1$ and $q_k = \alpha$. Now

$$\prod_{i=1}^{k-1} p_k = \prod_{i=1}^{k-1} 1 = 1, \tag{2.18}$$

and

$$\sum_{i=1}^{k-1} \left(q_i \Big/ \prod_{r=1}^{k-1} p_r \right) = \alpha \sum_{i=1}^{k-1} (1) = \alpha(k-1). \tag{2.19}$$

Therefore, the general solution, for $\beta = 1$, is

$$y_k = A + \alpha(k-1) = C + \alpha k, \tag{2.20}$$

where A and $C = A - \alpha$ are arbitrary constants.

2.2.3. Example C

The inhomogeneous equation

$$y_{k+1} - ky_k = 1 \tag{2.21}$$

has $p_k = k$ and $q_k = 1$. Since

$$\prod_{i=1}^{k} i = k!, \tag{2.22}$$

then

$$\sum_{i=1}^{k-1} \left(q_i \Big/ \prod_{r=1}^{i} p_r \right) = \sum_{i=1}^{k-1} \frac{1}{i!}, \tag{2.23}$$

and equation (2.21) has the solution

$$y_k = A(k-1)! + (k-1)! \sum_{i=1}^{k-1} \frac{1}{i!} \cdot \tag{2.24}$$

where A is an arbitrary constant.

2.2.4. Example D

Consider the homogeneous equation

$$y_{k+1} - e^{2k} y_k = 0, \tag{2.25}$$

where $p_k = e^{2k}$. Therefore,

$$\begin{aligned} y_k &= A \prod_{i=1}^{k-1} p_i = A \prod_{i=1}^{k-1} \exp(2i) \\ &= A \exp\left(2 \sum_{i=1}^{k-1} i\right) \cdot \end{aligned} \tag{2.26}$$

Using the fact that

$$\sum_{i=1}^{k-1} i = \frac{k(k-1)}{2} \tag{2.27}$$

gives

$$y_k = A \exp[k(k-1)]. \tag{2.28}$$

2.2.5. Example E

The equation

$$(k+1)y_{k+1} - ky_k = 0 \tag{2.29}$$

can be put in simpler form by the transformation

$$z_k = ky_k, \tag{2.30}$$

which gives

$$z_{k+1} - z_k = 0. \tag{2.31}$$

The solution to this latter equation is

$$z_k = A = \text{arbitrary constant.} \tag{2.32}$$

Therefore,

$$y_k = A/k. \tag{2.33}$$

Note that equation (2.29) corresponds to having $p_k = k/(k+1)$. Therefore,

$$\prod_{i=1}^{k-1} p_i = \prod_{i=1}^{k-1} \frac{i}{i+1} = \frac{(k-1)!}{k!} = \frac{1}{k}\cdot \tag{2.34}$$

Putting this result into equation (2.4) again gives the solution of equation (2.33).

2.2.6. Example F

The equation

$$(k+1)y_{k+1} - ky_k = k \tag{2.35}$$

can be written as

$$\Delta(ky_k) = k. \tag{2.36}$$

Therefore,

$$ky_k = \Delta^{-1}(k) = \tfrac{1}{2}k(k-1) + A, \tag{2.37}$$

where A is an arbitrary constant. Dividing by k gives

$$y_k = A/k + \tfrac{1}{2}(k-1). \tag{2.38}$$

Equation (2.35) can also be written as

$$y_{k+1} - \frac{k}{k+1} y_k = \frac{k}{k+1}\cdot \tag{2.39}$$

Therefore,

$$p_k = q_k = \frac{k}{k+1}, \tag{2.40}$$

and

$$\prod_{i=1}^{k-1} p_i = \frac{1}{k}, \tag{2.41}$$

$$\begin{aligned} \sum_{i=1}^{k-1} \left(q_i \Big/ \prod_{r=1}^{i} p_r \right) &= \sum_{i=1}^{k-1} \frac{i}{i+1}(i+1) \\ &= \sum_{i=1}^{k-1} i = \frac{k(k-1)}{2}. \end{aligned} \tag{2.42}$$

Substitution of equations (2.41) and (2.42) into equation (2.9) gives equation (2.38).

2.2.7. Example G

The inhomogeneous equation

$$y_{k+1} - y_k = e^k \tag{2.43}$$

has

$$p_k = 1, \qquad q_k = e^k. \tag{2.44}$$

Therefore,

$$\prod_{i=1}^{k-1} p_i = 1 \tag{2.45}$$

and

$$\sum_{i=1}^{k-1} \left(q_i \Big/ \prod_{r=1}^{i} p_r \right) = \sum_{i=1}^{k-1} e^i = \frac{e^k - e}{e - 1}. \tag{2.46}$$

Thus, the general solution of equation (2.43) is

$$y_k = A + \frac{e^k}{e-1}, \tag{2.47}$$

where A is an arbitrary constant.

2.3. $y_{k+1} - y_k = (n + 1)k^n$

The linear inhomogeneous difference equation

$$y_{k+1} - y_k = (n + 1)k^n \tag{2.48}$$

can be written as

$$\Delta y_k = (n + 1)k^n, \tag{2.49}$$

where n is an integer.

Consider the following equations and associated solutions:

$$\Delta y_k = 0, \qquad y_k = A; \tag{2.50}$$

$$\Delta y_k = 1, \qquad y_k = k + A; \tag{2.51}$$

$$\Delta y_k = 2k, \qquad y_k = k^2 - k + A; \tag{2.52}$$

$$\Delta y_k = 3k^2, \qquad y_k = k^3 - \tfrac{3}{2}k^2 + \tfrac{1}{2}k + A; \tag{2.53}$$

etc.

In each case, A is an arbitrary constant. For n arbitrary, we have

$$y_k = (n + 1)\Delta^{-1}(k^n) = (n + 1) \sum_{i=1}^{k-1} i^n + A, \tag{2.54}$$

which gives y_k as a polynomial of degree $(n + 1)$.

The Bernoulli polynomials $B_n(k)$ are defined to be solutions to the following difference equation:

$$B_n(k + 1) - B_n(k) = nk^{n-1}, \tag{2.55}$$

with a normalization for the arbitrary constant to be given below. These polynomials can be obtained from the following generating function:

$$G(k, \lambda) = \frac{\lambda e^{k\lambda}}{e^{\lambda} - 1}, \tag{2.56}$$

where

$$\lim_{\lambda \to 0} G(k, \lambda) = 1. \tag{2.57}$$

Expanding equation (2.56) in a power series in λ gives

$$G(k, \lambda) = \sum_{n=0}^{\infty} B_n(k) \frac{\lambda^n}{n!} \cdot \tag{2.58}$$

If we take the first differences of both sides of equation (2.58) with respect to k and equate the coefficients of the various powers of λ, we find that the $B_n(k)$ must satisfy equation (2.55). Thus, except for the value of the unknown constant A, the solutions of the difference equations, given by equations (2.50)–(2.53), are just the first three Bernoulli polynomials.

These normalization constants can be determined by noting that in each case $A = B_n(0)$. The $B_n(0)$ are known as the Bernoulli numbers. Their values follow from the relationships obtained by setting $k = 0$ in equations (2.56) and (2.57); that is,

$$\frac{\lambda}{e^{\lambda} - 1} = \left(1 + \frac{\lambda}{2!} + \frac{\lambda^2}{3!} + \cdots\right)^{-1} = \sum_{n=0}^{\infty} \frac{\lambda^n}{n!} B_n(0). \tag{2.59}$$

Comparison of the powers of λ gives

$$B_0(0) = 1, \quad B_1(0) = -\tfrac{1}{2}, \quad B_2(0) = \tfrac{1}{6}, \quad B_3(0) = 0,$$
$$B_4(0) = -\tfrac{1}{30}, \quad B_5(0) = 0, \quad B_6(0) = \tfrac{1}{42}, \ \ldots .$$

The corresponding Bernoulli polynomials are

$$\begin{aligned}
B_0(k) &= 1,\\
B_1(k) &= k - \tfrac{1}{2},\\
B_2(k) &= k^2 - k + \tfrac{1}{6},\\
B_3(k) &= k^3 - \tfrac{3}{2}k^2 + \tfrac{1}{2}k,\\
B_4(k) &= k^4 - 2k^3 + k^2 - \tfrac{1}{30},\\
B_5(k) &= k^5 - \tfrac{5}{2}k^4 + \tfrac{5}{3}k^3 - \tfrac{1}{6}k,\\
B_6(k) &= k^6 - 3k^5 + \tfrac{5}{2}k^4 - \tfrac{1}{2}k^2 + \tfrac{1}{42},\\
B_7(k) &= k^7 - \tfrac{7}{2}k^6 + \tfrac{7}{2}k^5 - \tfrac{7}{6}k^3 + \tfrac{1}{6}k,\\
B_8(k) &= k^8 - 4k^7 + \tfrac{14}{3}k^6 - \tfrac{7}{3}k^4 + \tfrac{2}{3}k^2 - \tfrac{1}{30},
\end{aligned}$$

etc.

It should be clear that the inhomogeneous difference equation

$$y_{k+1} - y_k = \sum_{m=0}^{n} a_m k^m, \tag{2.60}$$

where the a_m are given constants, can have its solution expressed in terms of Bernoulli polynomials.

Note that the Bernoulli polynomials are expressed as functions of k with $B_n(k)$ being of degree n. It is also possible to express k^n as a sum of Bernoulli polynomials. Using the results given above we obtain

$$\begin{aligned} k &= B_1(k) + \tfrac{1}{2}, \\ k^2 &= B_2(k) + B_1(k) + \tfrac{1}{3}, \\ k^3 &= B_3(k) + \tfrac{3}{2}B_2(k) + B_1(k) + \tfrac{1}{2}, \\ k^4 &= B_4(k) + 2B_3(k) + 2B_2(k) + B_1(k) + \tfrac{19}{30}, \end{aligned}$$

etc.

It can be shown that k^n has the following expansion:

$$k^n = \frac{1}{n+1} \sum_{i=0}^{n} \binom{n+1}{i} B_i(k). \tag{2.61}$$

2.3.1. Example

The equation

$$y_{k+1} - y_k = 1 - k + 2k^3 \tag{2.62}$$

has the particular solution

$$\begin{aligned} y_k &= \sum_{i=1}^{k-1} (1 - i + 2i^3) = \sum_{i=1}^{k-1} (1) - \sum_{i=1}^{k-1} i + 2 \sum_{i=1}^{k-1} i^3 \\ &= (k-1) - \frac{k(k-1)}{2} + \frac{(k-1)^2 k^2}{2}. \end{aligned} \tag{2.63}$$

The general solution is

$$y_k = \tfrac{1}{2}k^4 - k^3 + \tfrac{3}{2}k + A, \tag{2.64}$$

where A is an arbitrary constant. In terms of Bernoulli polynomials, this last expression reads

$$y_k = B_1(k) - \tfrac{1}{2}B_2(k) + \tfrac{1}{2}B_4(k) + A_1, \tag{2.65}$$

where A_1 is an arbitrary function.

The result of equation (2.65) could have been immediately written down by first noting that equation (2.60) is a linear equation and thus its particular solution is a sum of particular solutions of equations having the form

$$y_{k+1} - y_k = a_m k^m, \qquad 0 \leq m \leq n. \tag{2.66}$$

Under the transformation

$$y_k = \frac{a_m}{m+1} B_{m+1}(k), \tag{2.67}$$

equation (2.66) becomes

$$B_{m+1}(k+1) - B_{m+1}(k) = (m+1)k^m, \tag{2.68}$$

which is the defining difference equation for the Bernoulli polynomials. Therefore, from the known coefficients (a_m), the particular solution can be immediately obtained from equation (2.67).

2.4. $y_{k+1} = R_k y_k$

Let R_k be a rational function of k. It can be represented as follows:

$$R_k = \frac{C(k-\alpha_1)(k-\alpha_2)\cdots(k-\alpha_n)}{(k-\beta_1)(k-\beta_2)\cdots(k-\beta_m)}, \tag{2.69}$$

where C and the α's and β's are constants. Since

$$\Gamma(k+1-\alpha_i) = (k-\alpha_i)\Gamma(k-\alpha_i), \tag{2.70}$$

it is clear that the equation

$$y_{k+1} = R_k y_k \tag{2.71}$$

has the solution

$$y_k = AC^k \frac{\Gamma(k-\alpha_1)\Gamma(k-\alpha_2)\cdots\Gamma(k-\alpha_n)}{\Gamma(k-\beta_1)\Gamma(k-\beta_2)\cdots\Gamma(k-\beta_m)}, \tag{2.72}$$

where A is an arbitrary constant.

2.4.1. Example A

The equation

$$y_{k+1} = (k - k^2)y_k \tag{2.73}$$

can be written as

$$y_{k+1} = (-)k(k-1)y_k. \tag{2.74}$$

Its solution is

$$y_k = A(-)^k\Gamma(k)\Gamma(k-1) = A(-)^k(k-1)\Gamma^2(k-1), \tag{2.75}$$

where A is an arbitrary constant.

2.4.2. Example B

The equation

$$y_{k+1} = 3\frac{k+1}{k^2}y_k \tag{2.76}$$

has the solution

$$y_k = A3^k\frac{\Gamma(k+1)}{\Gamma^2(k)} = \mathrm{A}3^k\frac{k}{\Gamma(k)}, \tag{2.77}$$

where A is an arbitrary constant.

2.4.3. Example C

Consider the following equation:

$$\left[\frac{(k-1)(k+\frac{1}{2})}{k}\right]y_{k+1} = y_k. \tag{2.78}$$

The corresponding R_k is

$$R_k = \frac{k}{(k-1)(k+\frac{1}{2})}. \tag{2.79}$$

Therefore, the general solution of equation (2.78) is

$$y_k = A\frac{\Gamma(k)}{\Gamma(k-1)\Gamma(k+\frac{1}{2})}, \tag{2.80}$$

where A is an arbitrary constant.

2.4.4. Example D

The above technique can be used to obtain the general solution to first-order linear equations having linear coefficients. This class of equations take the form

$$(ak+b)y_{k+1} + (ck+d)y_k = 0, \tag{2.81}$$

where a, b, c, and d are constants. Using the fact that

$$\frac{ck+d}{ak+b} = \frac{c(k+d/c)}{a(k+b/a)}, \tag{2.82}$$

we can rewrite equation (2.81) to read

$$y_{k+1} + \frac{c}{a}\left[\frac{k+d/c}{k+b/a}\right]y_k = 0. \tag{2.83}$$

This last equation has the solution

$$y_k = A\left(-\frac{c}{a}\right)^k \frac{\Gamma(k+d/c)}{\Gamma(k+b/a)}, \tag{2.84}$$

where A is an arbitrary constant.

2.5. CONTINUED FRACTIONS

An interesting technique for obtaining a particular solution to equation (2.1) is the method of continued fractions. Equation (2.1) can be written as

$$a_k y_k = y_{k+1} + b_k, \tag{2.85a}$$

where $a_k = p_k$ and $b_k = -q_k$. Solving for y_k gives

$$y_k = \frac{b_k + y_{k+1}}{a_k}. \tag{2.85b}$$

A continued application of the result given by equation (2.85b) leads to a particular solution having the form of a continued fraction:

$$y_k = \cfrac{b_k + \cfrac{b_{k+1} + \cfrac{b_{k+2} + \cfrac{b_{k+3} + \cdot^{\cdot^{\cdot}}}{a_{k+3}}}{a_{k+2}}}{a_{k+1}}}{a_k}, \tag{2.86a}$$

which can be written as

$$y_k = \frac{b_k}{a_k} + \frac{b_{k+1}}{a_k a_{k+1}} + \frac{b_{k+2}}{a_k a_{k+1} a_{k+2}} + \cdots. \tag{2.86b}$$

The general solution is obtained by adding the solution of the homogeneous equation to the result given by equation (2.86b). It should be pointed out that the series solution of equation (2.86b) is, in general, an asymptotic series and, consequently, is in general not a convergence series.

2.5.1. Example A

Consider the equation

$$\lambda y_k = y_{k+1} - 1; \qquad \lambda = \text{constant}. \tag{2.87}$$

This corresponds to $a_k = \lambda$ and $b_k = -1$. Using any of the methods of this chapter, the general solution of equation (2.87) is found to be

$$y_k = A\lambda^k + \frac{1}{1-\lambda}, \tag{2.88}$$

where A is an arbitrary constant. Now according to equation (2.86b), a particular solution of equation (2.89) is

$$y_k = -\left(\frac{1}{\lambda}+\frac{1}{\lambda^2}+\frac{1}{\lambda^3}+\cdots\right). \tag{2.89}$$

If we use the result that for $|\lambda| > 1$, we have

$$\frac{1}{1-\lambda} = -\frac{1}{\lambda}\left[\frac{1}{1-1/\lambda}\right] = -\left(\frac{1}{\lambda}+\frac{1}{\lambda^2}+\frac{1}{\lambda^3}+\cdots\right), \tag{2.90}$$

then we can sum the series in λ on the right-hand side of equation (2.89) to obtain the particular solution given in equation (2.88).

2.5.2. Example B

The equation

$$y_{k+1} - \lambda y_k = k \tag{2.91}$$

has the general solution

$$y_k = A\lambda^k + \frac{k}{1-\lambda} - \frac{1}{(1-\lambda)^2}, \tag{2.92}$$

where A is an arbitrary constant. With this rewritten in the f

$$\lambda y_k = y_{k+1} - k, \tag{2.93}$$

we see that $a_k = \lambda$ *and* $b_k = -k$. Substitution of these results into equation (2.86b) gives the particular solution

$$y_k = -\left(\frac{k}{\lambda}+\frac{k+1}{\lambda^2}+\frac{k+2}{\lambda^3}+\cdots+\frac{k+n}{\lambda^{n+1}}+\cdots\right). \tag{2.94}$$

Now this last equation can be expressed as

$$\begin{aligned} y_k = &-k\left(\frac{1}{\lambda}+\frac{1}{\lambda^2}+\frac{1}{\lambda^3}+\cdots+\frac{1}{\lambda^n}+\cdots\right) \\ &-\frac{1}{\lambda^2}\left(1+\frac{2}{\lambda}+\frac{3}{\lambda^2}+\cdots+\frac{n+1}{\lambda^n}+\cdots\right). \end{aligned} \tag{2.95}$$

From the previous example, we know that the first sum can be replaced by $1/(1 - \lambda)$. Likewise, the second sum can be replaced by $1/(1 - \lambda)^2$ if we use the result ($|\lambda > 1$)

$$\begin{aligned}\frac{1}{(1-\lambda)^2} &= \frac{1}{\lambda^2(1-1/\lambda)^2} \\ &= \frac{1}{\lambda^2}\left(1+\frac{2}{\lambda}+\frac{3}{\lambda^2}+\cdots+\frac{n+1}{\lambda^n}+\cdots\right).\end{aligned} \tag{2.96}$$

Therefore, we have reproduced the particular solution of equation (2.91) as given by equation (2.95).

2.5.3. Example C

The equation

$$y_{k+1} - ky_k = -e^{-1}, \qquad a_k = k \quad \text{and} \quad b_k = e^{-1}, \tag{2.97}$$

has the particular solution

$$y_k = e^{-1} \sum_{n=0}^{\infty} \frac{1}{k(k+1)\cdots(k+n)}. \tag{2.98}$$

Equation (2.97) is a special case of

$$y_{k+1} - ky_k = -e^{-\rho}\rho^k, \tag{2.99}$$

where ρ is a constant. The general solution of this last equation is

$$y_k = A\Gamma(k) + e^{-\rho}\rho^k \sum_{n=0}^{\infty} \frac{\rho^n}{k(k+1)\cdots(k+n)}, \tag{2.100}$$

where A is an arbitrary constant. Equation (2.97) corresponds to the situation where $\rho = 1$.

Define $P(k;\ \rho)$ and $Q(k;\ \rho)$ as follows:

$$P(k;\ \rho) = e^{-\rho}\rho^k \sum_{n=0}^{\infty} \frac{\rho^n}{k(k+1)\cdots(k+n)}, \tag{2.101}$$

$$Q(k;\ \rho) = \Gamma(k) - P(k;\ \rho). \tag{2.102}$$

The special functions which are obtained when $\rho = 1$ are called Prym's functions.

Let us now consider $Q(z;\ \rho)$ and $P(z;\ \rho)$ as complex functions of the complex variable $k = z$. It can be shown that $Q(z;\ \rho)$ is an integral function, i.e., it has no singularities in the finite complex plane, while $P(z;\ \rho)$ is a meromorphic function of z in the whole plane with simple poles at zero and the negative integers.

2.6. $y_{k+1} - ky_k = P_k$

The difference equation

$$y_{k+1} - ky_k = P_k, \tag{2.103}$$

where P_k is a polynomial of degree n, can be solved in terms of the Prym functions introduced in Section 2.5.3.

To proceed, we express P_k in factorials:

$$P_k = \sum_{s=0}^{n} s!a_s \binom{k}{s}, \tag{2.104}$$

where the a_s are known coefficients. Now define the function f_k as follows:

$$f_k = \sum_{s=0}^{n-1} s!b_k \binom{k-1}{s}, \tag{2.105}$$

where, for the moment, the constant coefficients are unknown. However, we do wish to have the conditions

$$b_{-1} = b_n = 0 \tag{2.106}$$

satisfied. An easy calculation shows that

$$f_{k+1} - kf_k = \sum_{s=0}^{n} s!(b_s - b_{s-1}) \binom{k}{s}. \tag{2.107}$$

Now, let us choose the b_s such that

$$b_s - b_{s-1} = a_s, \qquad s = 1, 2, \ldots, n. \tag{2.108}$$

From this last relation, we have

$$
\begin{aligned}
b_1 &= a_1 + b_0, \\
b_2 &= a_1 + a_2 + b_0, \\
b_3 &= a_1 + a_2 + a_3 + b_0, \\
&\;\vdots \\
b_n &= a_1 + a_2 + \cdots + a_n + b_0.
\end{aligned}
\tag{2.109}
$$

The last of equations (2.108) and the condition given by equation (2.106) allow the conclusion

$$b_0 = -\sum_{s=1}^{n} a_s. \tag{2.110}$$

Since

$$b_s = \sum_{l=1}^{s} a_l + b_0, \tag{2.111}$$

we obtain

$$b_s = \sum_{l=1}^{s} a_l - \sum_{l=1}^{n} a_l = -\sum_{s+1}^{n} a_l. \tag{2.112}$$

These results allow a complete determination of f_k.

Now define the function v_k as follows:

$$y_k = v_k + f_k. \tag{2.113}$$

Substitution of equation (2.113) into equation (2.103) gives

$$v_{k+1} - kv_k = \sum_{s=0}^{n} a_s = \alpha. \tag{2.114}$$

Comparison of this expression with equation (2.97) allows us to conclude that a particular solution of equation (2.114) is

$$v_k = -e\alpha P(k;\ 1), \tag{2.115}$$

where $P(k;\ 1)$ is a Prym function. Therefore, the general solution of equation (2.114) is

$$v_k = A\Gamma(k) - e\alpha P(k;\ 1), \tag{2.116}$$

where A is an arbitrary constant. Finally, the general solution of equation (2.103) is

$$y_k = f_k + A\Gamma(k) - e\alpha P(k;\ 1). \tag{2.117}$$

2.7. A GENERAL FIRST-ORDER EQUATION: GEOMETRICAL METHODS

In this section, we will examine by means of geometrical techniques the possible solution behaviors of the difference equation

$$y_{k+1} = f(y_k), \tag{2.118}$$

where, in general, $f(y_k)$ is a nonlinear function of y_k. Our goal is to study the behavior of solutions of this equation for arbitrary initial values as $k \to \infty$.

Note that given any initial value y_0, equation (2.118) can be used to obtain y_1, which can then be used to obtain y_2, etc. Thus, to each initial value y_0, there corresponds a unique sequence $(y_0, y_1, y_2, y_3, \ldots)$. This procedure can be described graphically in the following manner:

(i) On rectangular axes plot y_{k+1} as a function of y_k using the functional relation given by equation (2.118). Also plot on this same graph the straight line $y_{k+1} = y_k$.

(ii) For any initial value y_0, draw a vertical line from the horizontal (y_k) axis, with this value, to the curve $y_{k+1} = f(y_k)$; this gives y_1.

(iii) To obtain the value y_2, draw a horizontal line over to the line $y_{k+1} = y_k$; this point, when projected on the y_k axis, gives the value y_2. This value y_2 can then be used, following the procedure given in (i) and (ii), to obtain y_3 and all the additional values of the sequence $(y_3, y_4, \ldots)$. The points where the curves $y_{k+1} = f(y_k)$ and $y_{k+1} = y_k$ intersect are the fixed points of equation (2.118); i.e., they correspond to constant solutions of equation (2.118). Therefore, the central question is whether the values of y_k converge to a fixed point $(y_k, y_{k+1}) = (a, a)$ as $k \to \infty$, or whether it oscillates or diverges to $\pm\infty$. An analysis of the details of this behavior will lead to useful asymptotic expansions for y_k. The particular procedures for obtaining these expressions will be presented in the next section.

The examples to follow illustrate the use of geometrical methods for obtaining information on the possible solution behaviors of equation (2.118).

2.7.1. Example A

Assume that the curve $y_{k+1} = f(y_k)$ does not intersect the line $y_{k+1} = y_k$. There are two cases to consider; see Figures 2–1 and 2–2.

For the first case, let the curve lie above the line $y_{k+1} = y_k$. In addition, we assume that the curve extends to (∞, ∞) in the first quadrant and $(-\infty, \infty)$ in the second quadrant or $(-\infty, -\infty)$ in the third quadrant.

Figure 2–1 shows the consequences of applying the above-stated geometrical procedure for two initial values indicated by y_0 and y_0'. Note that in each case as $k \to \infty$, $y_k \to \infty$. Thus, we conclude that if the curve $y_{k+1} = f(y_k)$ lies entirely above the line $y_{k+1} = y_k$, then y_k tends to $+\infty$.

Similarly, for the case where $y_{k+1} = f(y_k)$ lies entirely below $y_{k+1} = y_k$ and the curve extends from (∞, ∞) in the first quadrant or from $(\infty, -\infty)$ in the fourth quadrant to $(-\infty, -\infty)$ in the third quadrant, we conclude

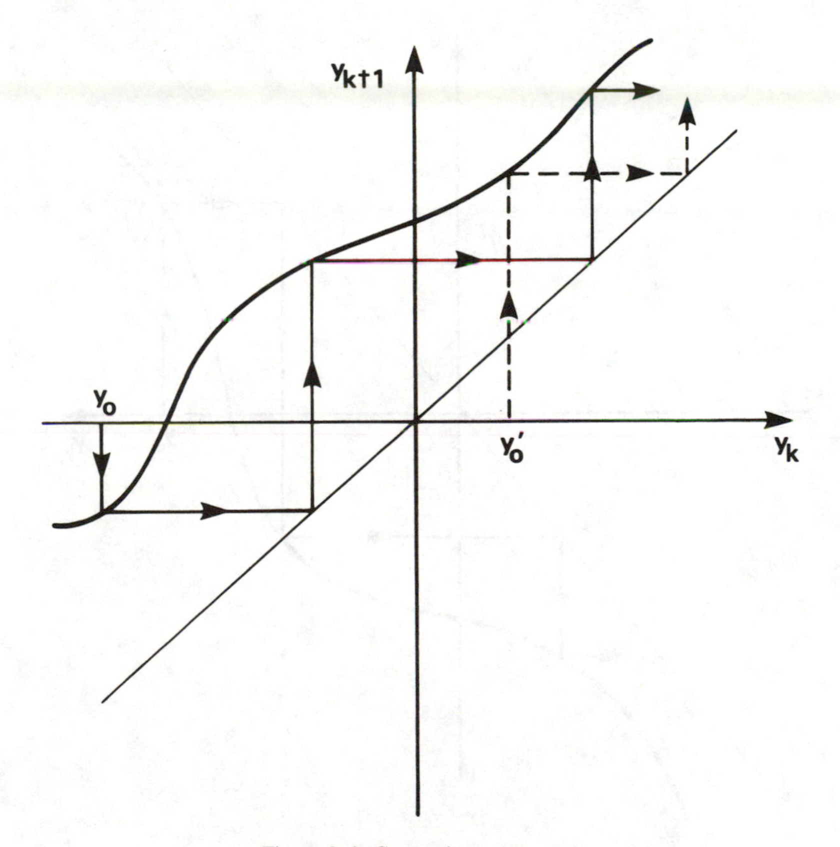

Figure 2–1. Curve above $y_{k+1} = y_k$.

that $y_k \rightarrow -\infty$ as $k \rightarrow \infty$. The corresponding geometrical construction is given in Figure 2–2.

2.7.2. Example B

We now consider the situation where the curve $y_{k+1} = f(y_k)$ intersects the line $y_{k+1} = y_k$ once. There are four separate cases to investigate; they are illustrated in Figure 2–3. (In each case, we assume that the branches of the curve extend to $\pm\infty$ in an appropriate fashion.)

Application of the geometrical procedures stated above shows that a stable fixed point is obtained only in (a). The other three cases have unstable fixed points.

In more detail, we have the following situation. Let $y_k = a$ be the fixed

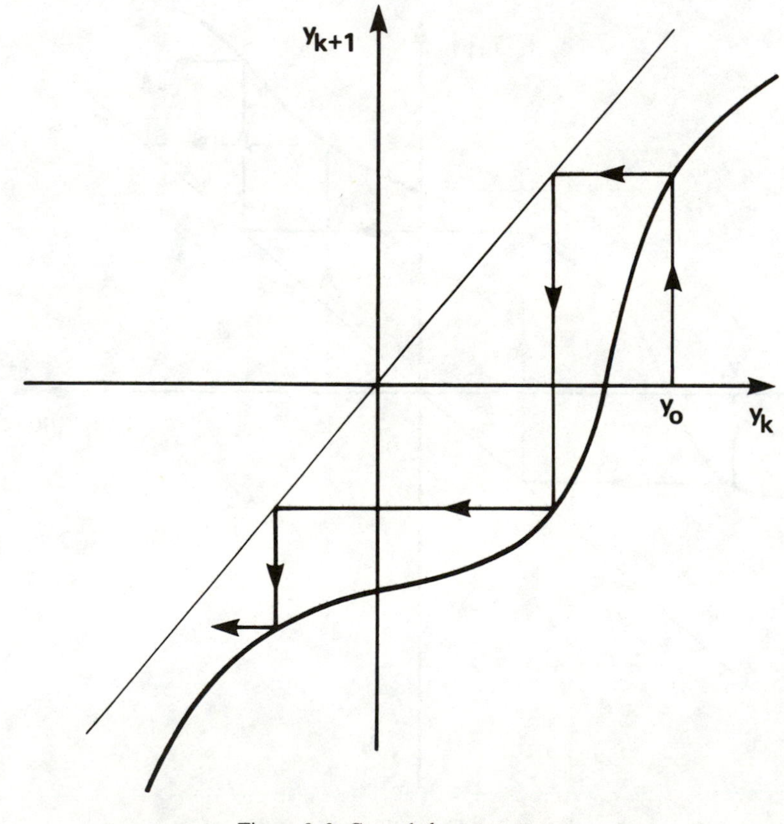

Figure 2–2. Curve below $y_{k+1} = y_k$.

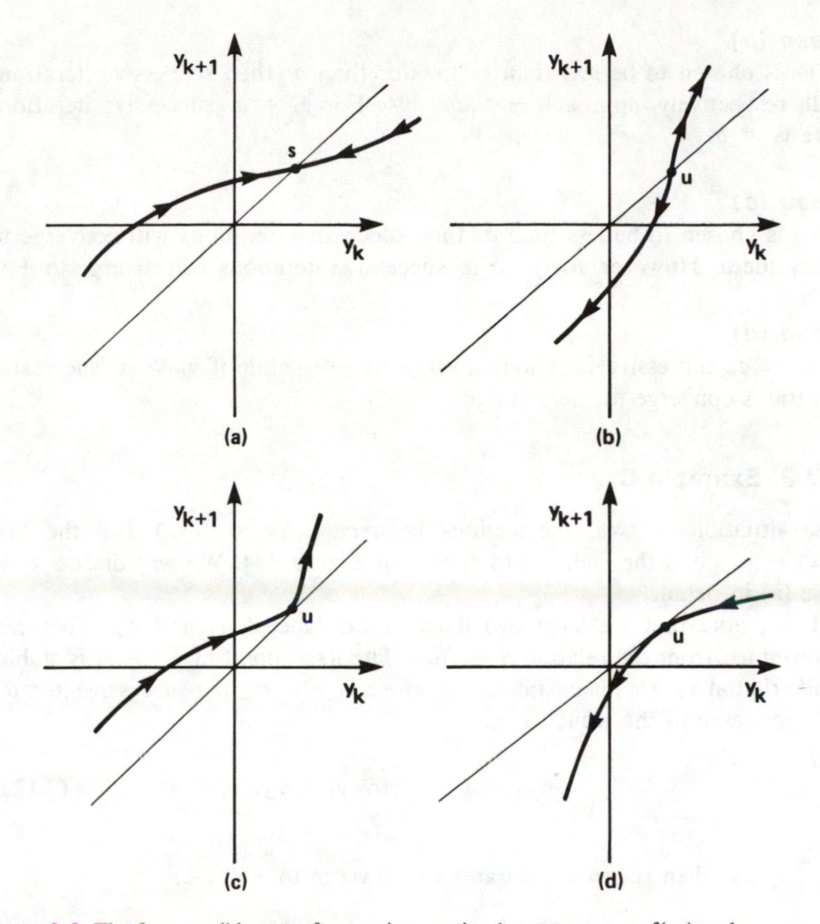

Figure 2–3. The four possible cases for one intersection between $y_{k+1} = f(y_k)$ and $y_{k+1} = y_k$. The stability of the fixed points is labeled by s = stable or u = unstable. The arrows indicate the direction of successive iterations. Cases (c) and (d) correspond to the curve having a common tangent with the line $y_{k+1} = y_k$ at the point of intersection.

point; it corresponds to the intersection of $y_{k+1} = f(y_k)$ and $y_{k+1} = y_k$. Its value is determined by the equation

$$a = f(a). \tag{2.119}$$

Case (a)

If y_0 is chosen arbitrarily, then successive iterations will converge to the value a, i.e.,

$$\lim_{k\to\infty} y_k = a, \qquad y_0 = \text{arbitrary}. \tag{2.120}$$

Case (b)

If y_0 is chosen to be less than or greater than a, then successive iterations will, respectively, approach $-\infty$ and $+\infty$. For $y_0 = a$, successive iterations give $y_k = a$.

Case (c)

If y_0 is chosen to be less than a, then successive iterations will converge to the value a. However, for $y_0 > a$, successive iterations will diverge to $+\infty$.

Case (d)

If $y_0 < a$, successive iterations diverge to $-\infty$, while if $y_0 > a$, successive iterations converge to the value a.

2.7.3. Example C

The situation of two intersections between $y_{k+1} = f(y_k)$ and the line $y_{k+1} = y_k$ gives the eight cases shown in Figure 2–4. We will discuss only case (a) in detail.

First, note that we have two fixed points labeled a_1 and a_2. They are determined from the relation $a = f(a)$. The fixed point at $y_k = a_1$ is stable, while that at $y_k = a_2$ is unstable. Therefore, if $y_0 < a_2$, the successive iterates will converge to the value a_1, *i.e.*,

$$\lim_{k\to\infty} y_k = a_1, \qquad \text{for } y_0 < a_2. \tag{2.121}$$

If $y_0 > a_2$, then successive iterates will diverge to $+\infty$, i.e.,

$$\lim_{k\to\infty} y_k = +\infty, \qquad \text{for } y_0 > a_2. \tag{2.122}$$

Note that in Figures 2–3 and 2–4 we have called a fixed point stable if there exists a neighborhood of the fixed point such that all initial values in this interval converge to the fixed point. Otherwise, the fixed point is unstable. Thus, for example, in both Figures 2–4(a) and 2–4(b), the fixed point a_1 is stable, while the fixed point a_2 is unstable.

2.8. A GENERAL FIRST-ORDER EQUATION: EXPANSION TECHNIQUES

We now investigate the construction of analytic approximations to the solutions of equation (2.118) in the neighborhood of a fixed point.

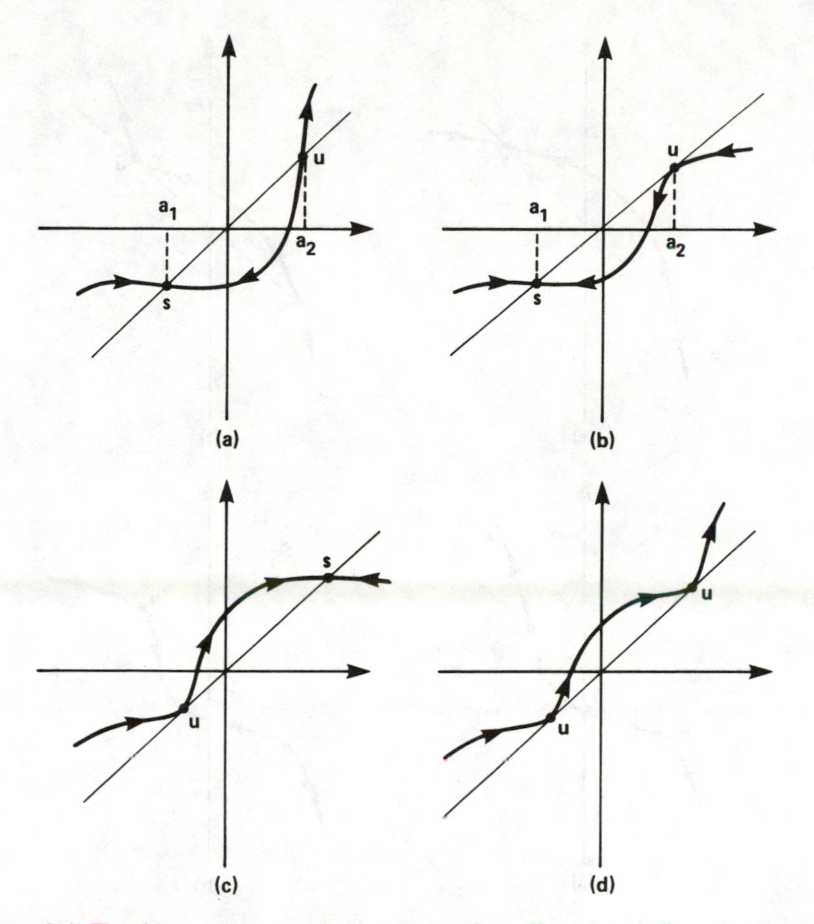

Figure 2–4. The eight possible cases for two intersections. The arrows indicate the direction of successive iterations. The stability of the fixed points is labeled by s = stable or u = unstable.

Let (a, a) be a fixed point of

$$y_{k+1} = f(y_k); \tag{2.123}$$

that is,

$$a = f(a). \tag{2.124}$$

Now let

$$y_k = a + u_k \tag{2.125}$$

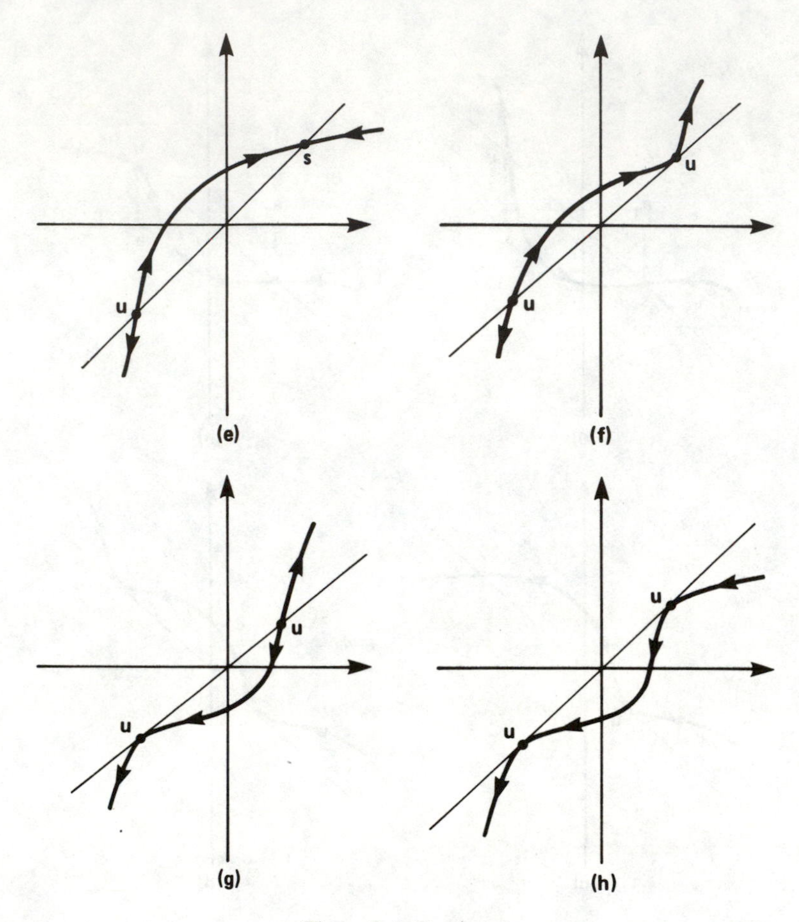

Figure 2–4. Continued.

and substitute this result into equation (2.123) and expand in a series in u_k; doing this gives

$$\begin{aligned} y_{k+1} &= f(a + u_k) \\ &= f(a) + f'(a)u_k + \tfrac{1}{2} f''(a)(u_k)^2 + \cdots . \end{aligned} \tag{2.126}$$

The first-order approximation is

$$y_{k+1} = a + f'(a)u_k. \tag{2.127}$$

For successive iterations to converge to a finite limit as $k \rightarrow \infty$, the following condition must be satisfied:

$$|y_{k+1} - a| < |y_k - a|, \tag{2.128}$$

or

$$|f'(a)u_k| < |u_k|, \tag{2.129}$$

which implies that

$$|f'(a)| < 1. \tag{2.130}$$

Therefore, we can conclude that if

$|f'(a)| < 1$, we have convergence,

$|f'(a)| = 1$, we have conditional convergence,

$|f'(a)| > 1$, we have divergence.

We now have enough information to construct asymptotic series for y_k at fixed points and at infinity. There are three distinct situations:

(i) $a = f(a)$ and $0 < |f'(a)| < 1$,
(ii) $a = f(a)$ and $f'(a) = 0$,
(iii) (y_k, y_{k+1}) approaching either (∞, ∞) or $(-\infty, -\infty)$.

(i) $a = f(a)$ and $0 < |f'(a)| < 1$

From equations (2.125) and (2.127), we have the first-order approximation

$$y_{k+1} - hy_k = (1 - h)a, \tag{2.131}$$

where $h = f'(a)$. Therefore,

$$y_k = a + Ah^k, \tag{2.132}$$

where A is an arbitrary constant. Let

$$t = Ah^k, \tag{2.133}$$

and define

$$z(t) = y_k. \tag{2.134}$$

Therefore,

$$y_{k+1} = a + ht, \tag{2.135}$$

and from $y_{k+1} = f(y_k)$, we have

$$a + ht = z(ht) = f[z(t)]. \tag{2.136}$$

Now assume that $z(t)$ has the asymptotic representation

$$z(t) = a + t + A_2 t^2 + A_2 t^3 + \cdots . \tag{2.137}$$

Consequently, we immediately obtain from equation (2.136) the following result:

$$\begin{aligned} a + ht + A_2 h^2 t^2 + A_3 h^3 t^3 + \cdots \\ = f(a) + f'(a)(t + A_2 t + A_3 t^3 + \cdots) + \tfrac{1}{2} f''(a)(t + A_2 t \\ + A_3 t^3 + \cdots)^2 + \cdots . \end{aligned} \tag{2.138}$$

Equating coefficients of corresponding powers of t gives

$$\begin{aligned} a &= f(a), \\ h &= f'(a), \\ A_2 h^2 &= \tfrac{1}{2}[f''(a) + 2A_2 f'(a)], \\ A_3 h^3 &= \tfrac{1}{6}[f'''(a) + 6A_2 f''(a) + 6A_3 f'(a)], \\ A_4 h^4 &= \tfrac{1}{24}[f''''(a) + 12A_2 f'''(a) + (24A_3 + 12A_2^2) f''(a) + 24A_4 f'(a)], \\ &\text{etc.} \end{aligned} \tag{2.139}$$

Solving for A_2, A_3, A_4, etc., gives

$$\begin{aligned} A_2 &= \frac{f''(a)}{2(h^2 - h)}, \\ A_3 &= \frac{(h^2 - h) f'''(a) + 3f''(a)^2}{6(h^3 - h)(h^2 - h)}, \\ A_4 &= \frac{f''''(a)}{24(h^4 - h)} + \frac{(3h + 5) f''(a) f'''(a)}{12(h^4 - h)(h^3 - h)} + \frac{(h + 5) f''(a)^3}{8(h^4 - h)(h^3 - h)(h^2 - h)}, \\ &\text{etc.} \end{aligned} \tag{2.140}$$

We conclude that for the difference equation $y_{k+1} = f(y_k)$, if there exists a real point (a, a) in the $y_k - y_{k+1}$ plane such that

$$a = f(a), \qquad |f'(a)| < 1, \tag{2.141}$$

then in the neighborhood of this point y_k has the asymptotic expansion

$$y_k = a + Ah^k + A_2A^2h^{2k} + A_3A^3h^{3k} + \cdots, \tag{2.142}$$

where A_2, A_3, etc., are given by the expressions of equation (2.140) and $h = f'(a)$.

(ii) $a = f(a)$ and $f'(a) = 0$

Let $f^{(n)}(a)$ be the first nonzero derivative and let $y_k = a + u_k$. Therefore, in the first approximation

$$y_{k+1} = f(a) + \frac{1}{n!} f^{(n)}(a)(u_k)^n, \tag{2.143}$$

and

$$u_{k+1} = \frac{1}{n!} f^{(n)}(a)(u_k)^n. \tag{2.144}$$

This last equation has the solution

$$u_k = \frac{A^{n^k}}{[f^{(n)}(a)/n!]^{1/(n-1)}}, \qquad |A| < 1, \tag{2.145}$$

where A is an arbitrary constant satisfying the condition guven by equation (2.145).

Let us now make the following definitions:

$$t = A^{n^k}, \qquad B_1 = \frac{1}{[f^{(n)}(a)/n!]^{1/(n-1)}}, \tag{2.146}$$

and assume that

$$y_k = z(t) = a + B_1t + B_2t_2 + \cdots. \tag{2.147}$$

Note that

$$t^n = A^{n(k+1)}. \tag{2.148}$$

Therefore, from $y_{k+1} = f(y_k)$, we have

$$z(t^n) = f[z(t)]. \tag{2.149}$$

On substituting the expansion of equation (2.147) into this expression, we obtain the result

$$\begin{aligned} a + B_1 t^n + B_2 t^{2n} + \cdots = f(a) + \frac{1}{n!} f^{(n)}(a)(B_1 t + B_2 t^2 \\ + B_3 t^3 + \cdots)^n + \frac{1}{(n+1)!} f^{(n+1)}(a)(B_1 t \\ + B_2 t^2 + B_3 t^3 + \cdots)^{n+1} + \cdots. \end{aligned} \tag{2.150}$$

Equating powers of t on each side allows the determination of B_2, B_3, etc.
For $n = 2$, we have

$$z(t^2) = f[z(t)], \tag{2.151}$$

$$a = f(a), \qquad f'(a) = 0, \tag{2.152}$$

and

$$\begin{aligned} B_1 &= \frac{2}{f''(a)}, \\ B_2 &= -\frac{2f'''(a)}{3f''(a)^3}, \\ B_3 &= \frac{5f'''(a)^3}{9f''(a)^5} - \frac{f''''(a)}{3f''(a)^4} - \frac{f'''(a)}{3f''(a)^3}, \end{aligned} \tag{2.153}$$

etc.

(iii) **(y_k, y_{k+1}) *Approaching Either* (∞, ∞) *or* $(-\infty, -\infty)$**

Let

$$y_{k+1} = p(y_k)^r, \tag{2.154}$$

where p and r are constants, and $r \geq 1$, be the desired asymptote to $y_{k+1} = f(y_k)$.

Consider the situation where

$$r = 1, \qquad p > 1. \tag{2.155}$$

The first approximation is

$$y_k = Ap^k, \qquad A = \text{arbitrary constant.} \tag{2.156}$$

Let

$$t = Ap^k, \qquad z(t) = y_k. \tag{2.157}$$

Therefore, $y_{k+1} = f(y_k)$ becomes

$$z(pt) = f[z(t)], \tag{2.158}$$

and the asymptotic expansion takes the form

$$z(t) = t + A_0 + \frac{A_1}{t} + \frac{A_2}{t^2} + \cdots. \tag{2.159}$$

Now let $r > 1$. Therefore, the first approximation is

$$y_k = p^{1/(1-r)} e^{Ar^k} = p^{1/(1-r)} C^{r^k}, \tag{2.160}$$

where A is an arbitrary constant and $C = e^A > 1$. With the definitions

$$t = C^{r^k}, \qquad z(t) = y_k, \tag{2.161}$$

the equation $y_{k+1} = f(y_k)$ becomes

$$z(t^r) = f[z(t)], \tag{2.162}$$

and corresponds to situation (ii) discussed above. Note that if r is a rational number $r = m/n$, where m and n are relatively prime integers, then by defining $t = s^n$, equation (2.163) takes the form

$$z(s^m) = f[z(s^n)]. \tag{2.163}$$

The following examples illustrate the application of the above techniques.

2.8.1. Example A

The equation

$$y_k y_{k+1} = 2y_k^2 + 1 \tag{2.164}$$

can be written as

$$y_{k+1} = 2y_k + 1/y_k. \tag{2.165}$$

In this form, it is easily seen that the curve in the (y_k, y_{k+1}) plane is a hyperbola with asymptotes $y_k = 0$ and $y_{k+1} = 2y_k$. There are no real fixed points since $a^2 = 2a^2 + 1$. Consequently, the curve does not intersect $y_{k+1} = y_k$. See Figure 2–5.

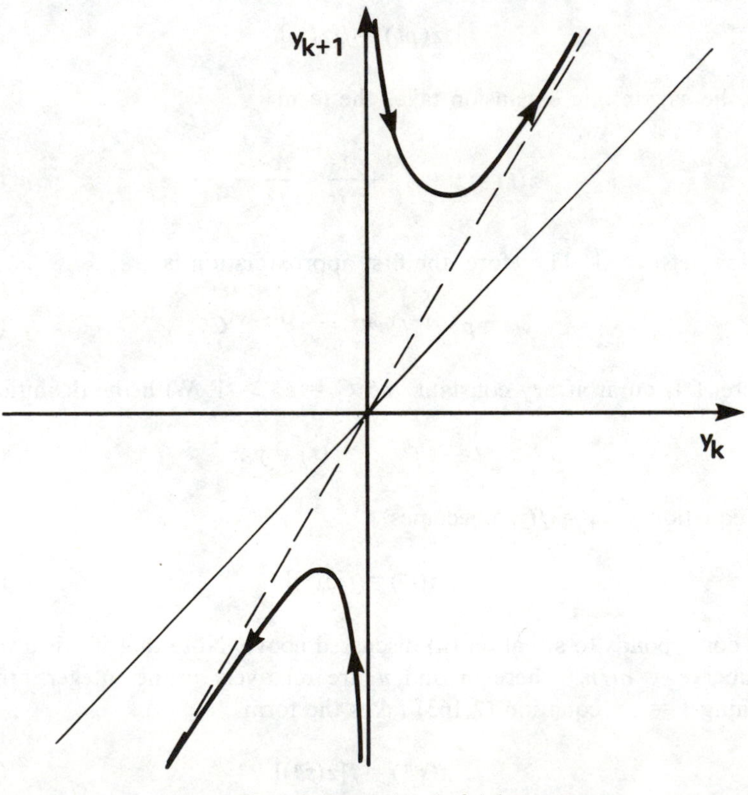

Figure 2–5. $y_k y_{k+1} = 2y_k^2 + 1$.

The asymptotic behavior is determined by

$$y_{k+1} = 2y_k, \tag{2.166}$$

the solution of which is

$$y_k = A2^k, \tag{2.167}$$

where A is an arbitrary constant. Therefore, if we let

$$t = A2^k \quad \text{and} \quad z(t) = y_k, \tag{2.168}$$

equation (2.164) becomes

$$z(t)z(2t) = 2z(t)^2 + 1, \tag{2.169}$$

where $z(t)$ has the expansion

$$z(t) = t + A_0 + \frac{A_1}{t} + \frac{A_2}{t^2} + \frac{A_3}{t^3} + \cdots. \tag{2.170}$$

Substitution of this result into equation (2.169) gives respectively, for the left- and right-hand sides

$$\begin{aligned} z(t)z(2t) = 2t^2 + 3A_0t + (A_0^2 + \tfrac{5}{2}A_1) + \frac{\tfrac{9}{4}A_2 + \tfrac{3}{2}A_0A_1}{t} \\ + \frac{\tfrac{17}{8}A_3 + \tfrac{5}{4}A_0A_2 + \tfrac{1}{2}A_1^2}{t^2} + \cdots \end{aligned} \tag{2.171}$$

and

$$\begin{aligned} 2z(t)^2 + 1 = 2t^2 + 4A_0t + (1 + 4A_1 + 2A_0^2) + \frac{4(A_2 + A_0A_1)}{t} \\ + \frac{2(2A_3 + 2A_0A_2 + A_1^2)}{t^2} + \cdots. \end{aligned} \tag{2.172}$$

Comparing the two expressions gives

$$
\begin{aligned}
3A_0 &= 4A_0, \\
A_0^2 + \tfrac{5}{2}A_1 &= 1 + 4A_1 + 2A_0^2, \\
\tfrac{9}{4}A_2 + \tfrac{3}{2}A_0A_1 &= 4(A_2 + A_0A_1), \\
\tfrac{17}{8}A_3 + \tfrac{5}{4}A_0A_2 + \tfrac{1}{2}A_1^2 &= 2(2A_3 + 2A_0A_2 + A_1^2),
\end{aligned}
\tag{2.173}
$$

which can be solved for A_0, A_1, A_2, A_3 to give

$$A_0 = 0, \qquad A_1 = -\tfrac{2}{3}, \qquad A_2 = 0, \qquad A_3 = -\tfrac{16}{45}. \tag{2.174}$$

Therefore,

$$z(t) = t - \frac{2}{3t} - \frac{16}{45t^3} - \cdots . \tag{2.175}$$

We conclude that equation (2.164) has the following asymptotic expansion at $k \to \infty$:

$$y_k = A2^k - \frac{2}{3}\left(\frac{2^{-k}}{A} + \frac{8}{15}\frac{2^{-3k}}{A^3} + \cdots\right). \tag{2.176}$$

2.8.2. Example B

Consider the following linear difference equation:

$$3y_{k+1} = y_k + 2. \tag{2.177}$$

This equation has a fixed point $y_k = 1$. The exact solution is given by the expression

$$y_k = 1 + A3^{-k}, \tag{2.178}$$

where A is an arbitrary constant. Consideration of both Figure 2–6 and the result of equation (2.178) shows that the fixed point is stable.

The linear difference equation

$$y_{k+1} = 2y_k - 1 \tag{2.179}$$

has a fixed point $y_k = 1$ and its exact solution is

$$y_k = 1 + A2^k, \tag{2.180}$$

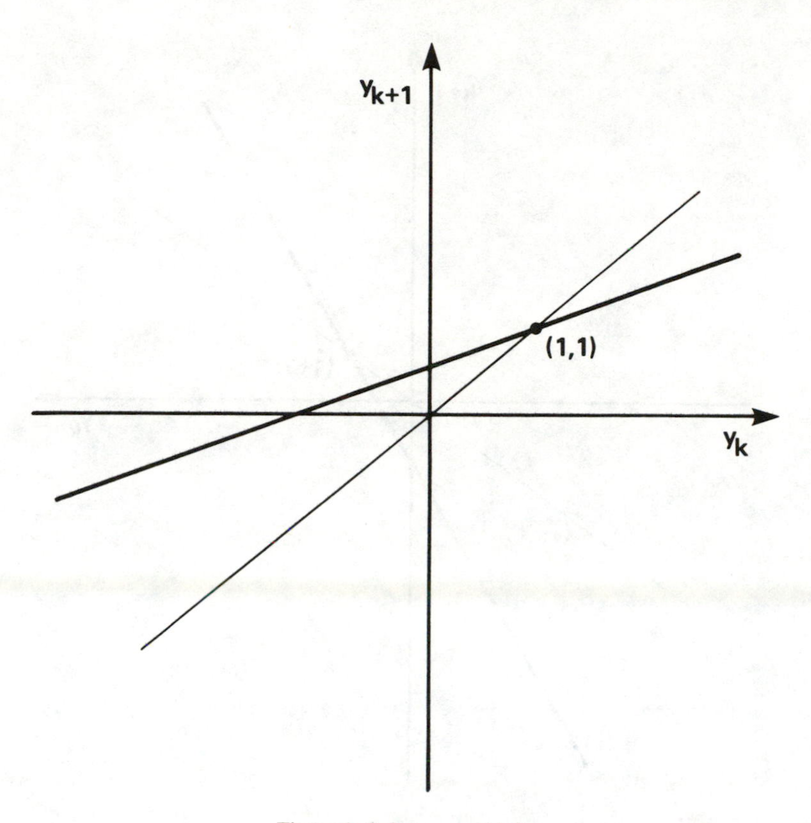

Figure 2–6. $3y_{k+1} = y_k + 2$.

where A is an arbitrary constant. For this case, the fixed point is unstable. See Figure 2–7.

Likewise, the equation

$$y_{k+1} = -2y_k + 3 \tag{2.181}$$

has the fixed point $y_k = 1$. Since the slope is larger in magnitude than one, the fixed point is unstable. The exact solution is

$$y_k = 1 + A(-2)^k, \tag{2.182}$$

where A is an arbitrary constant.

Note that for these three examples, we have, respectively, monotonic convergence, monotonic divergence, and oscillatory divergence. See Figures 2–6, 2–7, and 2–8.

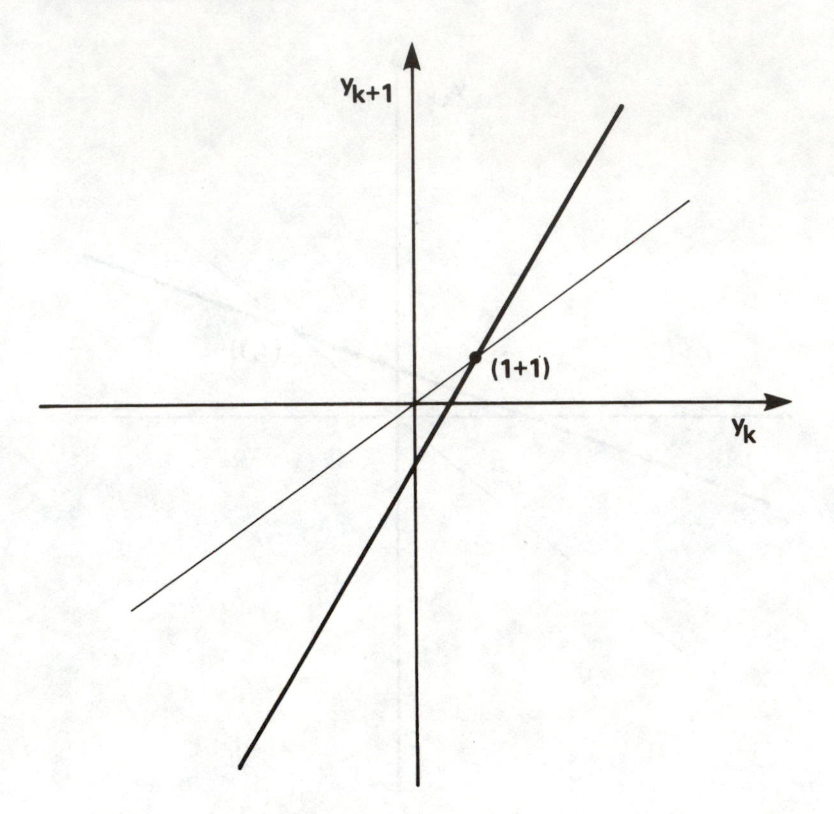

Figure 2–7. $y_{k+1} = 2y_k - 1$.

2.8.3. Example C

The equation

$$y_{k+1} = y_k + 1/y_k \tag{2.183}$$

does not cross the line $y_{k+1} = y_k$; see Figure 2–9. Therefore, we expect y_k to become unbounded as $k \to \infty$. However, we cannot use any of the above-discussed techniques to determine its asymptotic behavior.

Note that we cannot use the fact that $y_k \gg 1/y_k$ in equation (2.183) because the solution of the difference equation $y_{k+1} = y_k$ is a constant; this violates the condition that $y_k \to \infty$ as $k \to \infty$.

To proceed, we square both sides of equation (2.183) and obtain

$$y_{k+1}^2 = y_k^2 + 2 + 1/y_k^2. \tag{2.184}$$

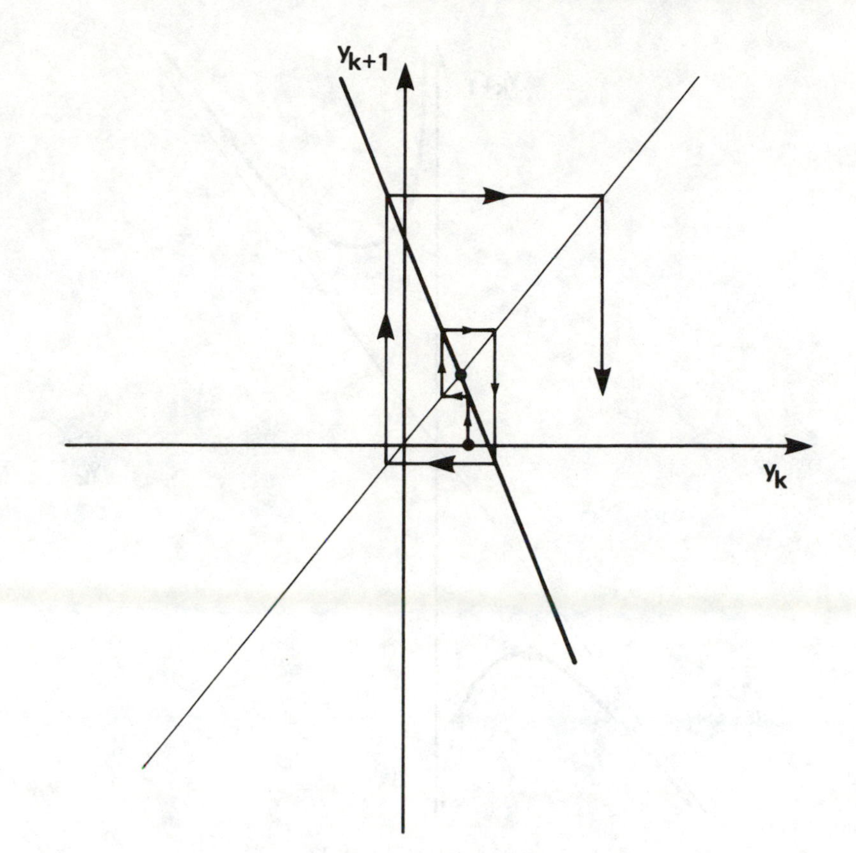

Figure 2–8. $y_{k+1} = -2y_k + 3$.

For this equation it is valid to neglect y_k^{-2} in comparison with 2 as $k \to \infty$. The resulting asymptotic difference equation

$$y_{k+1}^2 = y_k^2 + 2 \tag{2.185}$$

can be solved by making the transformation $v_k = y_k^2$. Its solution is

$$v_k = 2k, \qquad k \to \infty, \tag{2.186}$$

and

$$y_k = \sqrt{2k}, \qquad k \to \infty. \tag{2.187}$$

This is the dominant asymptotic behavior of y_k as $k \to \infty$.

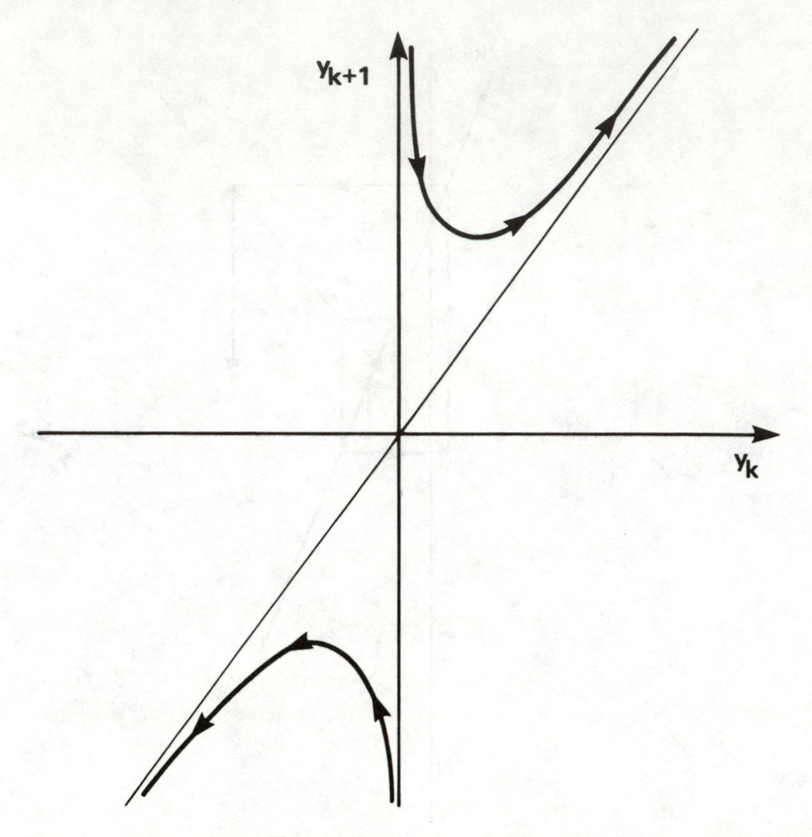

Figure 2–9. $y_{k+1} = y_k + 1/y_k$.

The next order correction to the result of equation (2.187) is obtained by letting

$$y_k = \sqrt{2k + \epsilon_k}, \tag{2.188}$$

where we assume that $\epsilon_k \ll 2k$ for $k \to \infty$. Substituting equation (2.188) into equation (2.184) and neglecting ϵ_k in comparison with $2k$ in the denominator of the fraction gives

$$\epsilon_{k+1} - \epsilon_k = 1/2k, \qquad k \to \infty, \tag{2.189}$$

the solution of which is

$$\epsilon_k = \tfrac{1}{2} \log k, \qquad k \to \infty. \tag{2.190}$$

From the result

$$\sqrt{2k+\epsilon_k} = \sqrt{2k}\left(1+\frac{\epsilon_k}{2k}\right)^{1/2} = \sqrt{2k}\left(1+\frac{\epsilon_k}{4k}+\cdots\right)$$
$$= \sqrt{2k} + \frac{1}{4}\frac{\log k}{\sqrt{2k}} + \cdots, \tag{2.191}$$

we conclude that the asymptotic behavior of equation (2.183) is

$$y_k = \sqrt{2k} + \frac{1}{4}\frac{\log k}{\sqrt{2k}} + \cdots. \tag{2.192}$$

2.8.4. Example D

The equation

$$y_{k+1} = \tfrac{1}{2}y_k + 2 - 3/2y_k \tag{2.193}$$

corresponds to a hyperbola in the (y_k, y_{k+1}) plane having asymptotes $y_k = 0$ and $y_{k+1} = \tfrac{1}{2}y_k + 2$. There are two fixed points, $y_k = 1$ and $y_k = 3$. The slopes at (1, 1) and (3, 3) are, respectively, 2 and $\tfrac{2}{3}$. Therefore, $y_k = 1$ is an unstable fixed point, while $y_k = 3$ is a stable fixed point.

A geometric analysis indicates that arbitrary starting points lead finally to the stable fixed point. See Figure 2–10.

The first approximation to the solution of equation (2.193) is given by the expression

$$y_k = 3 + A(\tfrac{2}{3})^k, \tag{2.194}$$

where A is an arbitrary constant. Therefore, if we let

$$t = A(\tfrac{2}{3})^k, \qquad z(t) = y_k, \tag{2.195}$$

then equation (2.193) becomes

$$2z(t)z\left(\frac{2t}{3}\right) = z(t)^2 + 4z(t) - 3. \tag{2.196}$$

We assume a solution of the form

$$z(t) = 3 + t + A_2t^2 + A_3t^3 + \cdots. \tag{2.197}$$

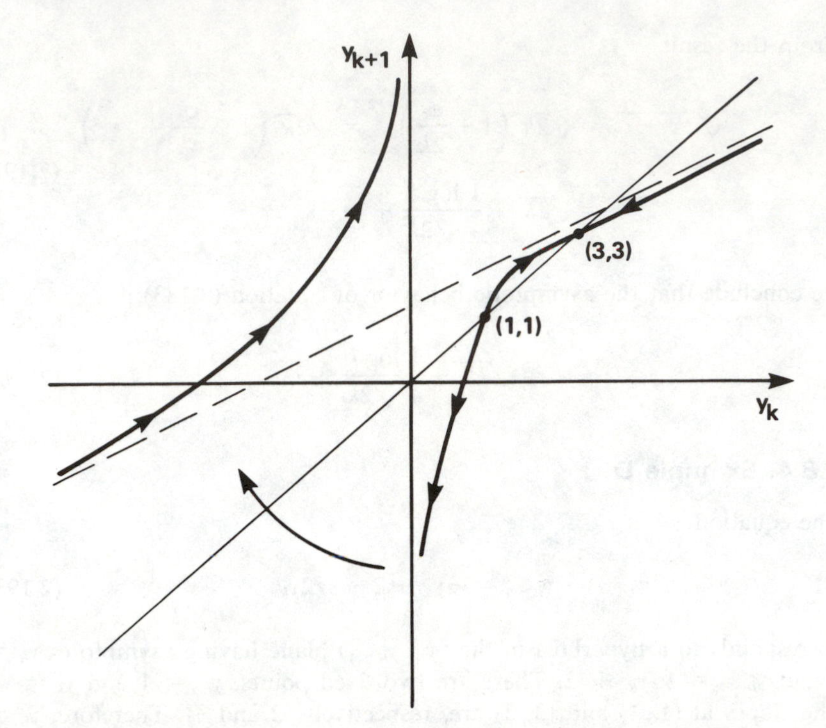

Figure 2–10. $2y_k y_{k+1} = y_k{}^2 + 4y_k - 3$. Note that initial values in the interval $0 < y_0 < 1$ go over to the branch in the left-hand plane.

Substitution of this expression into equation (2.197) gives, respectively, for the left- and right-hand sides

$$2z(t)z\left(\frac{2t}{3}\right) = 18 + 10t + \tfrac{2}{3}(13A_2 + 2)t^2 + \tfrac{2}{9}(35A_3 + 10A_2)t^2 + \cdots, \quad (2.198)$$

and

$$z(t)^2 + 4z(t) - 3 = 18 + 10t + (10A_2 + 1)t^2 + (10A_3 + 2A_2)t^3 + \cdots. \quad (2.199)$$

Equating the coefficients of corresponding powers of t gives

$$\begin{aligned} \tfrac{2}{3}(13A_2 + 2) &= 10A_2 + 1, \\ \tfrac{2}{9}(35A_3 + 10A_2) &= 10A_3 + 2A_2, \end{aligned} \quad (2.200)$$

and

$$A_2 = 1/4, \qquad A_3 = 1/40. \tag{2.201}$$

Therefore,

$$z(t) = 3 + t + 1/4 t^2 + 1/40 t^3 + \cdots,$$

and

$$y_k = 3 + A(2/3)^k + 1/4 A^2 (2/3)^{2k} + 1/40 A^3 (2/3)^{3k} + \cdots. \tag{2.202}$$

2.8.5. Example E

The graph of the nonlinear equation

$$y_{k+1}{}^2 = y_k \tag{2.203}$$

is shown in Figure 2–11. There are two fixed points, $y_k = 0$ and $y_k = 1$. The slopes of the curve at these fixed points are, respectively, ∞ and $1/2$; consequently, $y_k = 0$ is an unstable fixed point and $y_k = 1$ is a stable fixed point.

Note that the lower branch of the curve does not enter into consideration since it does not correspond to real values of y_k under continued iteration.

The transformation

$$v_k = \log y_k \tag{2.204}$$

will allow us to obtain the exact solution to equation (2.203). We find

$$2v_{k+1} = v_k, \tag{2.205}$$

the solution of which is

$$v_{k+1} = A(1/2)^k, \tag{2.206}$$

where A is an arbitrary constant. Therefore,

$$y_k = \exp[A(1/2)^k], \tag{2.207}$$

which has the asymptotic expansion

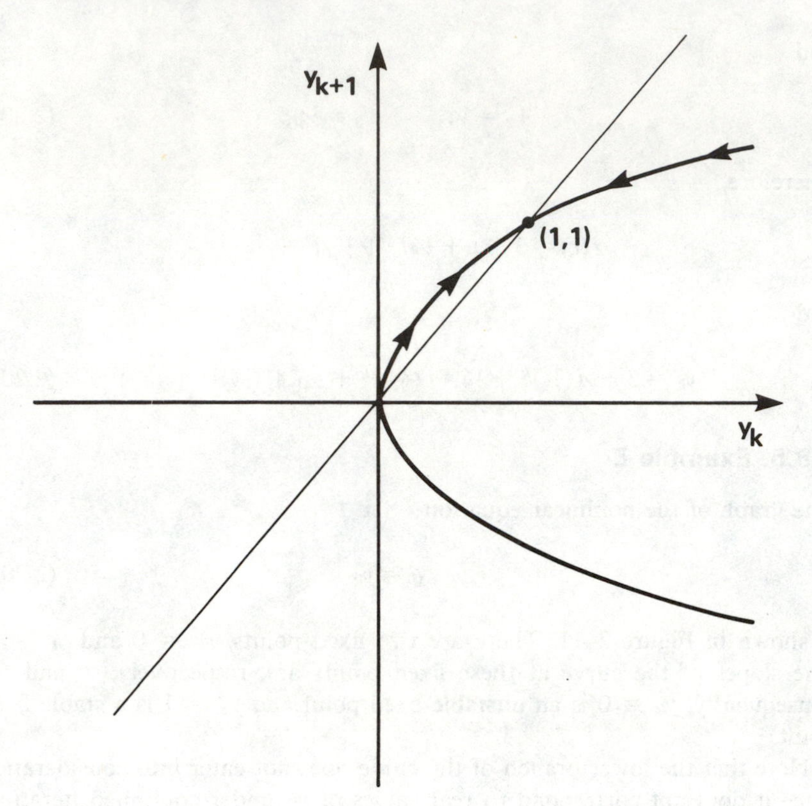

Figure 2–11. $y_{k+1}^2 = y_k$.

$$y_k = 1 + A(\tfrac{1}{2})^k + \tfrac{1}{2}A^2(\tfrac{1}{2})^{2k} + \tfrac{1}{6}A^3(\tfrac{1}{2})^{3k} + \cdots . \tag{2.208}$$

Rewrite equation (2.203) as follows:

$$y_{k+1} = \sqrt{y_k}, \tag{2.209}$$

and let

$$y_k = 1 + u_k, \tag{2.210}$$

where u_k is assumed to be small. Substitution of equation (2.210) into equation (2.209) gives

$$u_{k+1} = \tfrac{1}{2}u_k, \tag{2.211}$$

the solution of which is

$$u_k = A(\tfrac{1}{2})^k, \tag{2.212}$$

where A is an arbitrary constant. Therefore, the first approximation to y_k is

$$y_k = 1 + A(\tfrac{1}{2})^k. \tag{2.213}$$

Now let

$$t = A(\tfrac{1}{2})^k, \qquad z(t) = y_k. \tag{2.214}$$

Consequently, equation (2.203) becomes

$$z(\tfrac{1}{2}t)^2 = z(t). \tag{2.215}$$

Assuming a solution of the form

$$z(t) = 1 + t + A_2 t^2 + A_3 t^3 + \cdots \tag{2.216}$$

and substituting this result into equation (2.215) gives

$$z(\tfrac{1}{2}t)^2 = 1 + t + (\tfrac{1}{2}A_2 + \tfrac{1}{4})t^2 + (\tfrac{1}{4}A_3 + \tfrac{1}{4}A_2)t^3 + \cdots . \tag{2.217}$$

Comparison of equations (2.216) and (2.217) gives

$$\tfrac{1}{2}A_2 + \tfrac{1}{4} = A_2, \qquad \tfrac{1}{4}A_3 + \tfrac{1}{4}A_2 = A_3 \tag{2.218}$$

and

$$A_2 = \tfrac{1}{2}, \qquad A_3 = \tfrac{1}{6}. \tag{2,219}$$

Therefore,

$$z(t) = 1 + t + \tfrac{1}{2}t^2 + \tfrac{1}{6}t^3 + \cdots ,$$

which, on using the definition of t in equation (2.214), gives the same asymptotic expansion as equation (2.208).

3
LINEAR DIFFERENCE EQUATIONS

3.1. INTRODUCTION

Let the functions $a_0(k)$, $a_1(k)$, . . . , $a_n(k)$, and R_k be defined over a set of integers

$$k_1 \leq k \leq k_2, \tag{3.1}$$

where k_1 and k_2 can be either finite or unbounded in magnitude. An equation of the form

$$a_0(k)y_{k+n} + a_1(k)y_{k+n-1} + \cdots + a_n(k)y_k = R_k \tag{3.2}$$

is said to be linear. This equation is of order n if and only if

$$a_0(k)a_n(k) \neq 0, \tag{3.3}$$

for any k of equation (3.1). From the condition given by equation (3.3) we can, on division by $a_0(k)$ and relabeling the ratio of coefficient functions, write the general nth-order linear difference equation as follows:

$$y_{k+n} + a_1(k)y_{k+n-1} + \cdots + a_n(k)y_k = R_k. \tag{3.4}$$

Equation (3.4) is called homogeneous if R_k is identically zero for values of k in the interval given by equation (3.1); otherwise, it is called an inhomogeneous equation. In more detail,

$$y_{k+n} + a_1(k)y_{k+n-1} + \cdots + a_n(k)y_k = 0 \tag{3.5}$$

is an nth-order linear, homogeneous difference equation, while

$$y_{k+n} + a_1(k)y_{k+n-1} + \cdots + a_n(k)y_k = R_k \tag{3.6}$$

is an nth-order linear, inhomogeneous difference equation.

The following theorems are immediate consequences of the above definitions:

Theorem 3.1: Let c be an arbitrary constant; if y_k is a solution of equation (3.5), then cy_k is also a solution.

Proof: Multiply equation (3.5) by the constant c to obtain

$$c[y_{k+n} + a_1(k)y_{k+n-1} + \cdots + a_n(k)y] = 0. \tag{3.7}$$

This can be rewritten as

$$x_{k+n} + a_1(k)x_{k+n-1} + \cdots + a_n(k)x_k = 0, \tag{3.8}$$

where $x_k = cy_k$. Since equations (3.5) and (3.8) are the same, except for how we label the dependent variable, we conclude that if y_k is a solution to equation (3.5), then $x_k = cy_k$ is also a solution.

Theorem 3.2: Let c_1 and c_2 be arbitrary constants; let $y_k^{(1)}$ and $y_k^{(2)}$ be solutions of equation (3.5); then

$$y_k = c_1y_k^{(1)} + c_2y_k^{(2)} \tag{3.9}$$

is a solution.

Proof: This result follows directly from Theorem 3.1 and is called the principle of superposition.

Theorem 3.3: Let $y_k^{(1)}$ be a solution of equation (3.5) and let Y_k be a solution to equation (3.6); then

$$y_k = y_k^{(1)} + Y_k \tag{3.10}$$

is a solution to equation (3.6).

Proof: By assumption, we have

$$y_{k+n}^{(1)} + a_1(k)y_{k+n-1}^{(1)} + \cdots + a_n(k)y_k^{(1)} = 0 \tag{3.11}$$

and

$$Y_{k+n} + a_1(k)Y_{k+n-1} + \cdots + a_n(k)Y_k = R_k. \tag{3.12}$$

Adding these two equations gives

$$(y_{k+n}{}^{(1)} + Y_{k+n}) + a_1(k)(y_{k+n-1}{}^{(1)} + Y_{k+n}) + \cdots$$
$$+ a_n(k)(y_k{}^{(1)} + Y_k) = R_k. \quad (3.13)$$

Setting $y_k = y_k{}^{(1)} + Y_k$ gives the result of the theorem.

To proceed further, the following existence and uniqueness theorem is needed.

Theorem 3.4: There exists one, and only one, solution of equation (3.6) for which

$$y_k = A_0, \quad y_{k+1} = A_1, \; . \; . \; . \; , y_{k+n-1} = A_{n-1}, \quad (3.14)$$

where $(A_0, A_1, \; . \; . \; . \; , A_{n-1})$ are n arbitrary constants and k lies in the interval $k_1 \leq k \leq k_2$.

Proof: Equation (3.6) can be written

$$y_{k+n} = R_k - a_1(k)y_{k+n-1} - a_2(k)y_{k+n-2} + \cdots - a_n(k)y_k. \quad (3.15)$$

Consequently, y_{k_1+n} is uniquely determined by $y_{k_1}, y_{k_1+1}, \; . \; . \; . \; , y_{k_1+n-1}$. Likewise, y_{k_1+n+1} is uniquely determined by $y_{k_1+1}, y_{k_1+2}, \; . \; . \; . \; , y_{k_1+n}$, etc. Therefore, a unique solution can be obtained for all consecutive values of k in the interval $k_1 \leq k \leq k_2$.

3.1.1. Example A

The equation

$$(k+3)y_{k+1} + 5^k y_k = 0 \quad (3.16)$$

is a first-order, homogeneous difference equation. It is defined for all k except $k = -3$. The coefficient $a_1(k)$ is $5^k/(k+3)$.

The following equation is a second-order, inhomogeneous difference equation:

$$y_{k+2} - 6y_{k+1} + 3y_k = 1 + 2k^2 - 3 \cdot 2^k. \quad (3.17)$$

For this example, the coefficients are constants

$$a_1 = -6, \qquad a_2 = 3, \quad (3.18)$$

and the inhomogeneous term is

$$R_k = 1 + 2k^2 - 3 \cdot 2^k. \tag{3.19}$$

The equation

$$y_{k+4} - y_{k-4} = 5 \cos k + e^{-k} \tag{3.20}$$

is an eighth-order, inhomogeneous difference equation. It can be rewritten in the form

$$y_{k+8} - y_k = 5 \cos(k+8) + e^{-(k+8)}. \tag{3.21}$$

The coefficients are

$$\begin{gathered} a_1 = a_2 = a_3 = a_4 = a_5 = a_6 = a_7 = 0, \\ a_8 = -1, \end{gathered} \tag{3.22}$$

and the inhomogeneous term is

$$R_k = 5 \cos(k+8) + e^{-(k+8)}. \tag{3.23}$$

3.1.2. Example B

The second-order difference equation

$$y_{k+2} - 3y_{k+1} + 2y_k = 0 \tag{3.24}$$

has the two solutions $y_k^{(1)} = 1$ and $y_k^{(2)} = 2^k$, as can be shown by direct substitution into equation (3.24). For any two arbitrary constants, c_1 and c_2, it is easily seen that

$$y_k = c_1 y_k^{(1)} + c_2 y_k^{(2)} = c_1 + c_2 2^k \tag{3.25}$$

is also a solution, i.e.,

$$\begin{aligned} (c_1 + c_2 2^{k+2}) - 3(c_1 + c_2 2^{k+1}) + 2(c_1 + c_2 2^k) \\ = (1 - 3 + 2)c_1 + (4 - 6 + 2)c_2 2^k = 0. \end{aligned} \tag{3.26}$$

3.1.3. Example C

By direct substitution, we can show that the equation

$$y_{k+2} - 5y_{k+1} + 6y_k = 0 \tag{3.27}$$

has the solutions

$$y_k^{(1)} = 2^k, \qquad y_k^{(2)} = 3^k. \tag{3.28}$$

Therefore, for arbitrary constants c_1 and c_2,

$$y_k = c_1 2^k + c_2 3^k \tag{3.29}$$

is a solution to equation (3.27).

Likewise, we can show by direct substitution that

$$Y_k = \tfrac{1}{2}(k^2 + 3k + 5) \tag{3.30}$$

is a solution to the inhomogeneous equation

$$y_{k+2} - 5y_k + 6y_k = k^2. \tag{3.31}$$

Therefore, from Theorem 3.3, we conclude that

$$y_k = c_1 2^k + c_2 3^k + \tfrac{1}{2}(k^2 + 3k + 5) \tag{3.32}$$

is a solution to equation (3.31).

3.2. LINEARLY INDEPENDENT FUNCTIONS

A set of n functions $f_1(k), f_2(k), \ldots, f_n(k)$ is said to be linearly dependent, over the interval $k_1 \leq k \leq k_2$, if there exists a set of n constants $c_1, c_2, \ldots, c_n$, not all zero, such that

$$c_1 f_1(k) + c_2 f_2(k) + \cdots + c_n f_n(k) = 0. \tag{3.33}$$

If the set of functions $f_1(k), f_2(k), \ldots, f_n(k)$ is not linearly dependent, then the set is said to be linearly independent.

The Casoratian or Casorati determinant of n functions $f_1(k), f_2(k), \ldots, f_n(k)$ is defined as

$$C(k) = \begin{vmatrix} f_1(k) & f_2(k) & \cdots & f_n(k) \\ f_1(k+1) & f_2(k+1) & \cdots & f_n(k+1) \\ \cdot & \cdot & & \cdot \\ \cdot & \cdot & & \cdot \\ \cdot & \cdot & & \cdot \\ f_1(k+n-1) & f_2(k+n-1) & \cdots & f_n(k+n-1) \end{vmatrix} \tag{3.34}$$

The next two theorems show that the Casoratian plays an important role in determining whether particular sets of functions are linearly dependent or independent. (In the following theorems and for the remainder of this chapter, whenever we use the term "for all k," we mean the set of k values, $k_1 \leq k \leq k_2$, for which the functions are defined.)

Theorem 3.5: Let $f_1(k), f_2(k), \ldots, f_n(k)$ be n linearly dependent functions; their Casoratian equals zero for all k.

Proof: Since the n functions are linearly dependent, there exist constants, $c_1, c_2, \ldots, c_n$, not all zero, such that

$$c_1 f_1(k) + c_2 f_2(k) + \cdots + c_n f_n(k) = 0, \tag{3.35}$$

for all k. Therefore, we have

$$\begin{aligned} c_1 f_1(k) + c_2 f_2(k) + \cdots + c_n f_n(k) &= 0, \\ c_1 f_1(k+1) + c_2 f_2(k+1) + \cdots + c_n f_n(k+1) &= 0, \\ \cdot \qquad\qquad & \cdot \\ \cdot \qquad\qquad & \cdot \\ \cdot \qquad\qquad & \cdot \\ c_1 f_1(k+n-1) + c_2 f_2(k+n-1) + \cdots + c_n f_n(k+n-1) &= 0. \end{aligned} \tag{3.36}$$

Now assume that for $k = \bar{k}$ the Casoratian $C(\bar{k})$ is not equal to zero. Therefore, the only solution to equations (3.36) is

$$c_1 = c_2 = c_3 = \cdots = c_n = 0. \tag{3.37}$$

This contradicts our assumption that the c_i, $1 \leq i \leq n$, are not all simultaneously zero. Consequently, there can be no $k = \bar{k}$ such that $C(\bar{k}) = 0$.

Theorem 3.6: Let $f_1(k), f_2(k), \ldots, f_n(k)$ be n functions such that their Casoratian is zero for all k; then the n functions $f_1(k), f_2(k), \ldots, f_n(k)$ are linearly dependent.

Proof: At any value $k = \bar{k}$, the Casoratian $C(\bar{k})$ is zero. Therfore, the set of equations

$$\begin{aligned} c_1 f_1(\bar{k}) + c_2 f_2(\bar{k}) + \cdots + c_n f_n(\bar{k}) &= 0, \\ c_1 f_1(\bar{k}+1) + c_2 f_2(\bar{k}+1) + \cdots + c_n f_n(\bar{k}+1) &= 0, \\ &\vdots \\ c_1 f_1(\bar{k}+n-1) + c_2 f_2(\bar{k}+n-1) + \cdots + c_n f_n(\bar{k}+n-1) &= 0 \end{aligned} \tag{3.38}$$

has its associated determinant equal to zero. Therefore, we conclude that there exists a set of constants $c_1, c_2, \ldots, c_n$ which are not all zero and which also satisfy the equations (3.38). Consequently, the functions $f_i(k)$, $1 \leq i \leq n$, are linearly dependent.

The following examples will illustrate the use of these concepts.

3.2.1. Example A

Consider the three functions

$$f_1(k) = 3^k, \qquad f_2(k) = 3^{k+2}, \qquad f_3(k) = 2^k, \tag{3.39}$$

and form the linear combination

$$c_1 f_1(k) + c_2 f_2(k) + c_3 f_3(k) = (c_1 + 9c_2)3^k + c_3 2^k. \tag{3.40}$$

For the choice $c_1 = -9c_2$, where $c_2 \neq 0$ and $c_3 = 0$, the linear combination given by equation (3.40) is zero. Therefore, we conclude that the functions 3^k, 3^{k+2}, and 2^k are linearly dependent.

Note that the Casoratian for these functions is

$$C(k) = \begin{vmatrix} 3^k & 3^{k+2} & 2^k \\ 3^{k+1} & 3^{k+3} & 2^{k+1} \\ 3^{k+2} & 3^{k+4} & 2^{k+2} \end{vmatrix} = 0. \tag{3.41}$$

Again, we reach the conclusion that the above three functions are linearly dependent.

3.2.2. Example B

Consider the three functions

$$f_1(k) = 3^k, \qquad f_2(k) = k3^k, \qquad f_3(k) = k^2 3^k. \tag{3.42}$$

The linear combinations of these three functions can be written as follows:

$$c_1 f_1(k) + c_2 f_2(k) + c_3 f_3(k) = 3^k(c_1 + c_2 k + c_3 k^2). \tag{3.43}$$

Note that the right-hand side of equation (3.43) will be zero if and only if

$$c_1 + c_2 k + c_3 k^2 = 0. \tag{3.44}$$

But this can only occur for $c_1 = c_2 = c_3 = 0$. Therefore, the three functions given in equation (3.42) are linearly independent.

An easy calculation shows that the Casoratian for these functions is

$$C(k) = 2 \cdot 3^{3k+3}. \tag{3.45}$$

Since $C(k) \neq 0$, we again conclude that these three functions are linearly independent.

3.2.3. Example C

The three functions 2^k, 3^k, and $(-1)^k$ have the Casoratian

$$C(k) = \begin{vmatrix} 2^k & 3^k & (-1)^k \\ 2^{k+1} & 3^{k+1} & -(-1)^k \\ 2^{k+2} & 3^{k+2} & (-1)^k \end{vmatrix} = 12 \cdot 2^k \cdot 3^k \cdot (-1)^k \neq 0. \tag{3.46}$$

Therefore, the functions are linearly independent.

3.3. FUNDAMENTAL THEOREMS FOR HOMOGENEOUS EQUATIONS

One of the implications of Theorem 3.4 is that the linear, nth-order difference equation

$$y_{k+n} + a_1(k)y_{k+n-1} + \cdots + a_n(k)y_k = R_k \tag{3.47}$$

has a general solution that depends linearly on n arbitrary constants. We will now explore in some detail a number of important theorems concerning the solutions of the nth-order linear homogeneous equation

$$y_{k+n} + a_1(k)y_{k+n-1} + \cdots + a_n(k)y_k = 0. \tag{3.48}$$

Theorem 3.7: Let the functions $a_1(k)$, $a_2(k)$, . . . , $a_n(k)$ be defined for all k; let $a_n(k)$ be nonzero for all k; then there exist n linearly independent solutions $y_1(k)$, $y_2(k)$, (k), . . . , $y_n(k)$ of equation (3.48).

The proof will consist of two parts: first, we will explicitly construct n solutions; second, we will show these solutions to be linearly independent.

Proof: From Theorem 3.4, we know that a unique solution will be obtained if we specify any n consecutive values of y_k. Thus, to determine the n solutions $y_i(k)$, $1 \leq i \leq n$, we need to define their values at n consecutive values of k. Therefore, pick some value for k, say $k = \bar{k}$, and define the n functions $\hat{y}_i(k)$ as follows on the interval $\bar{k} \leq \bar{k} \leq \bar{k} + n - 1$:

$$\hat{y}_i(k) = \delta_{k,\bar{k}+i-1}, \qquad 1 \leq i \leq n, \tag{3.49}$$

where the delta function is

$$\delta_{l,k} = \begin{cases} 0, & \text{for } l \neq k, \\ 1, & \text{for } l = k, \end{cases} \tag{3.50}$$

The substitution of these $\hat{y}_i(t)$ into equation (3.48) allows the determination of the various $\hat{y}_i(t)$ for all values of k. Thus, from the existence Theorem 3.4, these functions are unique and we have constructed n different solutions to equation (3.48).

Let us now show that the n solutions $\hat{y}_i(k)$ are linearly independent. We will use the method of contradiction.

Assume that the $\hat{y}_i(k)$, $1 \leq i \leq n$, are linearly dependent. Therefore, there exist constants c_i, $1 \leq i \leq n$, not all zero, such that

$$c_1\hat{y}_1(k) + c_2\hat{y}_2(k) + \cdots + c_n\hat{y}_n(k) = 0. \tag{3.51}$$

From this last relationship, it follows that

$$
\begin{aligned}
c_1\hat{y}_1(k) + c_2\hat{y}_2(k) + \cdots + c_n\hat{y}_n(k) &= 0, \\
c_1\hat{y}_1(k+1)c_2\hat{y}_2(k+1) + \cdots + c_n\hat{y}_n(k+1) &= 0, \\
\vdots \qquad & \qquad \vdots \\
c_1\hat{y}_1(k+n-1) + c_2\hat{y}_2(k+n-1) + \cdots + c_n\hat{y}_n(k+n-1) &= 0.
\end{aligned}
\tag{3.52}
$$

Now equations (3.52) are a set of n linear equations in the n unknowns c_i, $1 \leq i \leq n$. Furthermore, the determinant associated with equations (3.52) is just the Casoratian of the n functions $y_i(k)$, $1 \leq i \leq n$:

$$
C(k) = \begin{vmatrix}
\hat{y}_1(k) & \hat{y}_2(k) & \cdots & \hat{y}_n(k) \\
\hat{y}_1(k+1) & \hat{y}_2(k+1) & \cdots & \hat{y}_n(k+1) \\
\vdots & \vdots & & \vdots \\
\hat{y}_1(k+n-1) & \hat{y}_2(k+n-1) & \cdots & \hat{y}_n(k+n-1)
\end{vmatrix} \cdot \tag{3.53}
$$

Let us now calculate the value of the Casoratian for $k = \bar{k}$. From the defining equations for the $\hat{y}_i(k)$, equation (3.50), this determinant has one for all its diagonal elements and zero for all other elements. Therefore, $C(k) = 1$ for $k = \bar{k}$. We conclude, since the determinant of the system of equations is nonzero, that the only values of the c_i, $1 \leq i \leq n$, that satisfy equations (3.52) are $c_i = 0$. However, this contradicts the assumption that the solutions $\hat{y}_i(k)$, $1 \leq i \leq n$, are linearly dependent; therefore, they must be linearly independent.

Defintion: A fundamental set of solutions of equation (3.48) is any n functions $y_i(k)$, $1 \leq i \leq n$, which are solutions of equation (3.48) and whose Casoratian, $C(k)$, is nonzero for all k.

The next theorem shows that a fundamental set of solutions exists for equation (3.48).

Theorem 3.8: Let $\hat{y}_i(k)$, $1 \leq i \leq n$, be the n linearly independent solutions of equation (3.48) as defined by equation (3.49). These solutions are a fundamental set of solutions.

Proof: Our task is to show that the Casoratian, $C(k)$, of the functions $\hat{y}_i(k)$ is nonzero for all k. First, we have shown that $C(k)$ is nonzero for $k = \bar{k}$. From equation (3.53), we have

$$C(k+1) = \begin{vmatrix} \hat{y}_1(k+1) & \hat{y}_2(k+1) & \cdots & \hat{y}_n(k+1) \\ \hat{y}_1(k+2) & \hat{y}_2(k+2) & \cdots & \hat{y}_n(k+2) \\ \cdot & \cdot & & \cdot \\ \cdot & \cdot & & \cdot \\ \cdot & \cdot & & \cdot \\ \hat{y}_1(k+n) & \hat{y}_2(k+n) & \cdots & \hat{y}_n(k+n) \end{vmatrix}. \tag{3.54}$$

Since the $\hat{y}_i(k)$ are solutions of equation (3.48), we have

$$\hat{y}_n(k+n) = -[a_1(k)\hat{y}_i(k+n-1) + \cdots + a_n(k)\hat{y}_i(k)]. \tag{3.55}$$

If the result of equation (3.55) is substituted into equation (3.54) and $a_1(t)$ times row $n-1$ is subtracted from row n, $a_2(t)$ times row $n-2$ is subtracted from row n, etc., then we obtain

$$C(k+1) = -a_n(k) \begin{vmatrix} \hat{y}_1(k+1) & \hat{y}_2(k+1) & \cdots & \hat{y}_n(k+1) \\ \hat{y}_1(k+2) & \hat{y}_2(k+2) & \cdots & \hat{y}_n(k+2) \\ \cdot & \cdot & & \cdot \\ \cdot & \cdot & & \cdot \\ \cdot & \cdot & & \cdot \\ \hat{y}_1(k+n-1) & \hat{y}_2(k+n-1) & \cdots & \hat{y}_n(k+n-1) \\ \hat{y}_1(k) & \hat{y}_2(k) & \cdots & \hat{y}_n(k) \end{vmatrix}. \tag{3.56}$$

Rearranging the rows and comparing with equation (3.53) gives

$$C(k+1) = (-1)^n a_n(k) C(k). \tag{3.57}$$

Since $C(k) \neq 0$, for $k = \bar{k}$, and since $a_n(k)$ is nonzero for all k, we conclude that $C(k)$ is nonzero for all k. Therefore, the $\hat{y}_i(k)$, $1 \leq i \leq n$, make up a fundamental set of solutions.

Theorem 3.9: Every solution y_k of equation (3.48) can be written as a linear combination of the functions $\hat{y}_i(k)$, $1 \leq i \leq n$, as defined by equation (3.49).

Proof: Let

$$Y_k = c_1\hat{y}_1(k) + c_2\hat{y}_2(k) + \cdots + c_n\hat{y}_n(k), \tag{3.58}$$

for constants c_i, $1 \le i \le n$. For $k = \bar{k}$, we have $\hat{y}_1(k) = 1$ and $\hat{y}_i(k) = 0$ for $i \ne 1$. Define $c_1 = y_k$ for $k = \bar{k}$.

Next, let $k = \bar{k} + 1$; we now have $\hat{y}_2(k) = 1$ and $\hat{y}_i(k) = 0$ for $i \ne 2$. Define $c_2 = y_k$ for $k = \bar{k} + 1$.

Continuing this process, we define $c_i = y_k$ for $k = \bar{k} + i - 1$.

If we now substitute these values of c_i, $1 \le i \le n$, into equation (3.58), we find that $Y_k = y_k$ for $k = \bar{k}, \bar{k} + 1, \ldots, \bar{k} + n - 1$. Since $Y_k = y_k$ for n consecutive values of k, we conclude, from Theorem 3.4, that $Y_k = y_k$ for all k. Therefore, any solution of equation (3.48) can be written as a linear combination of the functions $\hat{y}_i(k)$, $1 \le i \le n$.

Theorem 3.10: Every fundamental set of solutions of equation (3.48) is linearly independent.

Proof: Let $y_i(k)$, $1 \le i \le n$, be a fundamental set of solutions. If we assume that they are linearly dependent, then there exist constants c_i, $1 \le i \le n$, not all zero, such that

$$c_1y_1(k) + c_2y_2(k) + \cdots + c_ny_n(k) = 0. \tag{3.59}$$

Therefore,

$$\begin{aligned}
c_1y_1(k) + c_2y_2(k) + \cdots + c_ny_n(k) &= 0,\\
c_1y_1(k+1) + c_2y_2(k+1) + \cdots + c_ny_n(k+1) &= 0,\\
&\ \ \vdots\\
c_1y_1(k+n-1) + c_2y_2(k+n-1) + \cdots + c_ny_n(k+n-1) &= 0.
\end{aligned} \tag{3.60}$$

Since the $y_i(k)$, $1 \le i \le n$, are a fundamental set of solutions, the Casoratian $C(k)$ is nonzero for all k. However, the Casoratian is also equal to the determinant of the linear system of equations for the c_i given by equation (3.60). Therefore, we conclude that the only solution is $c_i = 0$, $1 \le i \le n$. This contradicts the assumption that the $y_i(k)$ are linearly dependent; consequently, they are linearly independent.

Theorem 3.11: An nth-order linear difference equation has n and only n linearly independent solutions.

Proof: This theorem follows from the previous theorems.

Theorem 3.12: The general solution of equation (3.48) is given by

$$y_k = c_1 y_1(k) + c_2 y_2(k) + \cdots + c_n y_n(k), \tag{3.61}$$

where the c_i, $1 \leq i \leq n$, are n arbitrary constants and the $y_i(k)$, $1 \leq i \leq n$, are a fundamental set of solutions.

Proof: This theorem follows directly from the definition of a fundamental set of solutions and Theorems 3.8, 3.9, and 3.10.

Before we finish this section, it should be pointed out that, except for a constant, the Casoratian for the n linearly independent solutions of an nth-order linear difference equation can be determined from a knowledge of the coefficient $a_n(k)$; see equations (3.48) and (3.57). We have

$$C(k+1) = (-1)^n a_n(k) C(k); \tag{3.62}$$

therefore, the Casoratian satisfies a first-order linear difference equation whose solution is

$$C(k) = (-1)^{nk} C(\bar{k}) \prod_{i=\bar{k}}^{k-1} a_n(i). \tag{3.63}$$

3.4. INHOMOGENEOUS EQUATIONS

In this section, we present a method for obtaining a particular solution Y_k to the inhomogeneous equation (3.6) given that we know the solution to the homogeneous equation (3.5). The technique to be developed is called the method of variation of constants.

To proceed, let $y_k^{(i)}$, $1 \leq i \leq n$, be a fundamental set of solutions of the homogeneous equation (3.5). We want to determine functions $C_i(k)$, $1 \leq i \leq n$, such that

$$Y_k = C_1(k) y_k^{(1)} + C_2(k) y_k^{(2)} + \cdots + C_n(k) y_k^{(n)} \tag{3.64}$$

is a particular solution to the inhomogeneous equation (3.6). Now, we have

$$Y_{k+1} = \sum_{i=1}^{n} C_i(k+1) y_{k+1}^{(i)}, \tag{3.65}$$

and

$$C_i(k+1) = C_i(k) + \Delta C_i(k). \tag{3.66}$$

Therefore,

$$Y_{k+1} = \sum_{i=1}^{n} C_i(k) y_{k+1}^{(i)}, \tag{3.67}$$

if we set

$$\sum_{i=1}^{n} y_{k+1}^{(i)} \Delta C_i(k) = 0. \tag{3.68}$$

Likewise, we have from equation (3.67)

$$Y_{k+2} = \sum_{i=1}^{n} C_i(k+1) y_{k+2}^{(i)} \tag{3.69}$$

and

$$Y_{k+2} = \sum_{i=1}^{n} C_i(k) y_{k+2}^{(i)}, \tag{3.70}$$

if we set

$$\sum_{i=1}^{n} y_{k+2}^{(i)} \Delta C_i(k) = 0. \tag{3.71}$$

Continuing this process, we obtain

$$Y_{k+n-1} = \sum_{i=1}^{n} C_i(k) y_{k+n-1}^{(i)}, \tag{3.72}$$

$$\sum_{i=1}^{n} y_{k+n-1}^{(i)} \Delta C_i(k) = 0, \tag{3.73}$$

and

$$Y_{k+n} = \sum_{i=1}^{n} C_i(k) y_{k+n}^{(i)} + \sum_{i=1}^{n} y_{k+n}^{(i)} \Delta C_i(k). \tag{3.74}$$

Substituting equations (3.67), (3.70)–(3.72), and (3.74) into the inhomogeneous equation (3.6) gives

$$\sum_{i=1}^{n} y_{k+n}^{(i)} \Delta C_i(k) = R_k. \tag{3.75}$$

Note that equations (3.68), (3.71)–(3.73), and (3.75) are a set of n linear equations for the $\Delta C_i(k)$, $1 \leq i \leq n$. Solving gives

$$\Delta C_i(\mathrm{k}) = \frac{f_i(k)}{C(k+1)}, \tag{3.76}$$

where $C(k+1)$ is the Casoratian of the fundamental set of solutions to the homogeneous equation; consequently, $C(k+1) \neq 0$; and the functions $f_i(k)$, $1 \leq i \leq n$, are known functions given in terms of R_k and the $y_k^{(i)}$. The equations (3.76) are linear, first-order equations and can be immediately solved to obtain the functions $C_i(k)$, $1 \leq i \leq n$.

3.4.1. Example A

An easy calculation shows that

$$y_k^{(1)} = 2^k, \qquad y_k^{(2)} = 1 \tag{3.77}$$

are solutions to the homogeneous equation

$$y_{k+2} - 3y_{k+1} + 2y_k = 0. \tag{3.78}$$

Furthermore, their Casoratian is given by the following expression:

$$C(k+1) = \begin{vmatrix} 2^{k+1} & 1 \\ 2^{k+2} & 1 \end{vmatrix} = -2^{k+1}. \tag{3.79}$$

Let us use these results to calculate a particular solution to the inhomogeneous equation

$$y_{k+2} - 3y_{k+1} + 2y_k = 4^k + 3k^2. \tag{3.80}$$

Let Y_k be the particular solution, where

$$Y_k = C_1(k)2^k + C_2(k). \tag{3.81}$$

The functions $C_1(k)$ and $C_2(k)$ satisfy the following equations:

$$\begin{aligned} 2^{k+1}\Delta C_1(k) + \Delta C_2(k) &= 0, \\ 2^{k+2}\Delta C_1(k) + \Delta C_2(k) &= 4^k + 3k^2. \end{aligned} \tag{3.82}$$

Solving for $\Delta C_1(k)$ and $\Delta C_2(k)$ gives

$$\begin{aligned} \Delta C_1(k) &= 2^{k-1} + 3 \cdot (\tfrac{1}{2})^{k+1} k^2, \\ \Delta C_2(k) &= -4^k - 3k^2. \end{aligned} \tag{3.83}$$

Therefore,

$$\begin{aligned} C_1(k) &= A + \tfrac{1}{2} \sum_{i=0}^{k-1} 2^i + \tfrac{3}{2} \sum_{i=0}^{k-1} i^2 (\tfrac{1}{2})^i \\ &= A - \tfrac{1}{2}(1 - 2^k) - 6[(\tfrac{1}{2}k^2 + k + \tfrac{3}{2})(\tfrac{1}{2})^k - \tfrac{3}{2}] \\ &= A_1 + 2^k - (3k^2 + 6k + 9)(\tfrac{1}{2})^k \end{aligned} \tag{3.84}$$

and

$$\begin{aligned} C_2(k) &= B - \sum_{i=0}^{k-1} 4^i - 3 \sum_{i=0}^{k-1} i^2 \\ &= B_1 - \tfrac{1}{3}4^k - \left(k^3 - \tfrac{3}{2}k^2 + \frac{k}{2}\right), \end{aligned} \tag{3.85}$$

where A and B are arbitrary constants and

$$A_1 = A - \tfrac{1}{2} + 9, \qquad B_1 = B + \tfrac{1}{3}. \tag{3.86}$$

Now the particular solution should not contain any terms with arbitrary constants or terms which are proportional to the solutions of the homogeneous equation. Consequently, if the results of equations (3.84) and (3.85) are substituted into equation (3.81), we obtain

$$Y_k = C_1 2^k + C_2 + \tfrac{1}{6}4^k - k^3 - \tfrac{3}{2}k^2 - \tfrac{13}{2}k, \tag{3.87}$$

where $C_1 = A_1$ and $C_2 = B_1 - 9$.

The first two terms contain arbitrary constants and are proportional to the fundamental set of solutions of the homogeneous equation. Therefore, the particular solution to equation (3.80) is

$$Y_k = \tfrac{1}{6}4^k - k^3 - \tfrac{3}{2}k^2 - \tfrac{13}{2}k. \tag{3.88}$$

Note, however, that the full expression on the right-hand side of equation (3.87) is the general solution of equation (3.80).

3.4.2. Example B

Consider the inhomogeneous equation

$$y_{k+1} - 2y_k \cos\phi + y_{k-1} = R_k, \qquad \sin\phi \neq 0, \tag{3.89}$$

where ϕ is a constant and R_k is an arbitrary function of k. The fundamental set of solutions to the homogeneous equation is

$$y_k^{(1)} = \cos(k\phi), \qquad y_k^{(2)} = \sin(k\phi). \tag{3.90}$$

The unknown functions in the particular solution expression

$$Y_k = C_1(k)\cos(k\phi) + C_2(k)\sin(k\phi) \tag{3.91}$$

satisfy the following equations:

$$\begin{aligned} [\cos(k+1)\phi]\Delta C_1(k) + [\sin(k+1)\phi]\Delta C_2(k) &= 0, \\ [\cos(k+2)\phi]\Delta C_1(k) + [\sin(k+2)\phi]\Delta C_2(k) &= R_{k+1}. \end{aligned} \tag{3.92}$$

Solving for $\Delta C_1(k)$ and $\Delta C_2(k)$ gives

$$\begin{aligned} \Delta C_1(k) &= -\frac{R_{k+1}\sin(k+1)\phi}{\sin\phi}, \\ \Delta C_2(k) &= \frac{R_{k+1}\cos(k+1)\phi}{\sin\phi}. \end{aligned} \tag{3.93}$$

Therefore,

$$\begin{aligned} C_1(k) &= C_1 - \sum_{i=1}^{k} \frac{R_i \sin(i\phi)}{\sin\phi}, \\ C_2(k) &= C_2 - \sum_{i=1}^{k} \frac{R_i \cos(i\phi)}{\sin\phi}, \end{aligned} \tag{3.94}$$

where C_1 and C_2 are arbitrary constants. Substitution of equations (3.94) into equation (3.91) and retaining the arbitrary constants gives the general solution to equation (3.89):

$$y_k = C_1 \cos(k\phi) + C_2 \cos(k\phi) - \frac{\cos(k\phi)}{\sin\phi} \sum_{i=1}^{k} R_i \sin(i\phi) + \frac{\sin(k\phi)}{\sin\phi} \sum_{i=1}^{k} R_i \cos(i\phi). \quad (3.95)$$

Using trigonometric substitution, this latter equation can be written

$$y_k = C_1 \cos(k\phi) + C_2 \sin(k\phi) + \frac{1}{\sin\phi} \sum_{i=1}^{k} R_i \sin(k-i)\phi. \quad (3.96)$$

Example C

The second-order, inhomogeneous equation

$$(k+4)y_{k+2} + y_{k+1} - (k+1)y_k = 1 \quad (3.97)$$

has the following two solutions:

$$y_k^{(1)} = \frac{1}{(k+1)(k+2)}, \quad (3.98)$$

$$y_k^{(2)} = \frac{(-1)^{k+1}(2k+3)}{4(k+1)(k+2)} \quad (3.99)$$

to its associated homogeneous equation

$$(k+4)y_{k+2} + y_{k+1} - (k+1)y_k = 0. \quad (3.100)$$

These functions have the Casoratian

$$C(k+1) = \frac{(-1)^{k+1}}{(k+2)(k+3)(k+4)}. \quad (3.101)$$

The particular solution takes the form

$$Y_k = c_1(k)y_k^{(1)} + c_2(k)y_k^{(2)}. \quad (3.102)$$

Direct calculation shows that $c_1(k)$ and $c_2(k)$ satisfy the equations

$$\begin{aligned} \Delta c_1(k) &= \tfrac{1}{4}(2k+5), \\ \Delta c_2(k) &= (-1)^{k+1}. \end{aligned} \tag{3.103}$$

Summing these expressions gives

$$c_1(k) = \tfrac{1}{4} \sum_{i=0}^{k} (2i+5) + c_1 = \tfrac{1}{4}(k+1)^2 + c_1 \tag{3.104}$$

and

$$c_2(k) = -\sum_{i=0}^{k} (-1) = -\tfrac{1}{2}[1+(-1)^k] + c_2, \tag{3.105}$$

where c_1 and c_2 are arbitrary constants. Substituting equations (3.98), (3.99), (3.104), and (3.105) into equation (3.102) and dropping the terms that contain the arbitrary constants gives

$$Y_k = \frac{k+1}{4(k+2)} + \frac{(2k+3)[1+(-1)^k]}{8(k+1)(k+2)}. \tag{3.106}$$

Now the term $(2k+3)(-)^k/8(k+1)(k+2)$ can be dropped, since it is proportional to the solution $y_k^{(2)}$ of the homogeneous equation. Combining the two expressions on the right-hand side of equation (3.106) gives

$$Y_k = \frac{k^2+3k+\tfrac{5}{2}}{4(k+1)(k+2)}. \tag{3.107}$$

The term associated with the constant in the numerator can also be dropped, since it is proportional to $y_k^{(1)}$. Therefore, we have the following expression for the particular solution of equation (3.97):

$$y_k = \frac{k(k+3)}{4(k+1)(k+2)}. \tag{3.108}$$

3.5. SECOND-ORDER EQUATIONS

In this section, we consider the general second-order linear difference equation. A number of techniques are described that allow the determination of the general solution if one solution is known to the homogeneous equation. We indicate, where appropriate, how the results can be generalized to higher-order equations.

Consider the second-order, homogeneous equation

$$y_{k+2} + p_k y_{k+1} + q_k y_k = 0, \tag{3.109}$$

where p_k and q_k are given functions of k. Suppose that one solution, $y_k^{(1)}$, is known for equation (3.109). We now show that a second solution, $y_k^{(2)}$, can be found. To proceed, we note that the Casoratian $C(k)$ satisfies the equation

$$C(k+1) = q_k C(k). \tag{3.110}$$

Therefore,

$$C(k) = AQ_k = A \prod_{i=1}^{k-1} q_i, \tag{3.111}$$

where A is an arbitrary, nonzero constant. Now

$$\frac{C(k)}{y_k^{(1)} y_{k+1}^{(1)}} = \frac{y_k^{(1)} y_{k+1}^{(2)} - y_k^{(2)} y_{k+1}^{(1)}}{y_k^{(1)} y_{k+1}^{(1)}} = \Delta \frac{y_k^{(2)}}{y_k^{(1)}}. \tag{3.112}$$

Applying Δ^{-1} to both sides gives

$$y_k^{(2)} = y_k^{(1)} \Delta^{-1} \frac{C(k)}{y_k^{(1)} y_{k+1}^{(1)}} = A y_k^{(1)} \Delta^{-1} \frac{Q_k}{y_k^{(1)} y_{k+1}^{(1)}}. \tag{3.113}$$

Thus, if a solution $y_k^{(1)}$ is known to equation (3.109), a second, linearly independent solution can be found and is given by the expression in equation (3.113). [Note that the constant A in equation (3.113) can be dropped.] The general solution to equation (3.109) is

$$y_k = c_1 y_k^{(1)} + c_2 y_k^{(2)}, \tag{3.114}$$

where c_1 and c_2 are arbitrary constants.

Consider now the inhomogeneous equation

$$y_{k+2} + p_k y_{k+1} + q_k y_k = R_k, \tag{3.115}$$

where p_k, q_k, and R_k are given functions of k. We now show that if one solution, $y_k^{(1)}$, is known to the homogeneous equation

$$y_{k+2} + p_k y_{k+1} + q_k y_k = 0, \qquad (3.116)$$

then the general solution to equation (3.115) can be determined.

Assume that the general solution to equation (3.115) can be written

$$y_k = y_k^{(1)} u_k, \qquad (3.117)$$

where $y_k^{(1)}$ is a solution to equation (3.116) and u_k is an unknown function. Substitution of equation (3.117) into equation (3.115) gives

$$y_{k+2}^{(1)} u_{k+2} + p_k y_{k+1}^{(1)} u_{k+1} + q_k y_k^{(1)} u_k = R_k. \qquad (3.118)$$

Now multiply equation (3.116) by u_{k+1} and subtract the result from equation (3.119); doing this gives

$$y_{k+2}^{(1)}(u_{k+2} - u_{k+1}) - q_k y_k^{(1)}(u_{k+1} - u_k) = R_k, \qquad (3.119)$$

or

$$y_{k+2}^{(1)} \Delta u_{k+1} - q_k y_k^{(1)} \Delta u_k = R_k. \qquad (3.120)$$

Let $x_k = \Delta u_k$; therefore

$$y_{k+2}^{(1)} x_{k+1} - q_k y_k^{(1)} x_k = R_k. \qquad (3.121)$$

This latter equation is a first-order, inhomogeneous equation and has the solution

$$x_k = AP_k + P_k \sum_{i=1}^{k-1} \frac{R_i}{P_{i+1}}, \qquad (3.122)$$

where A is an arbitrary constant and

$$P_k = \prod_{i=1}^{k-1} \frac{q_i y_i^{(1)}}{y_{i+1}^{(1)}}. \qquad (3.123)$$

Replacing x_k by $u_{k+1} - u_k$ again gives a first-order, inhomogeneous equation; its solution is

$$u_k = A\Delta^{-1} P_k + \Delta^{-1}\left(P_k \sum_{i=1}^{k-1} \frac{R_i}{P_{i+1}}\right) + B, \qquad (3.124)$$

where B is an arbitrary constant. If equation (3.124) is substituted into equation (3.117), then the following result is obtained:

$$y_k = Ay_k^{(1)}\,\Delta^{-1}P_k + By_k^{(1)} + y_k^{(1)}\,\Delta^{-1}\left(P_k \sum_{i=1}^{k-1} \frac{R_i}{P_{i+1}}\right). \tag{3.125}$$

Note that this expression contains two arbitrary constants, A and B, as is required for it to be the general solution to equation (3.115). In equation (3.125), the first two terms on the right-hand side correspond to two linearly independent solutions to the homogeneous equation (3.116), while the third term is the particular solution to the inhomogeneous equation (3.115).

Let us consider again equations (3.115) and (3.117). Using the definition of u_k, given by equation (3.117), we can write equation (3.119) as follows:

$$y_{k+2}^{(1)}\left(\frac{y_{k+2}}{y_{k+2}^{(1)}} - \frac{y_{k+1}}{y_{k+1}^{(1)}}\right) - q_k y_k^{(1)}\left(\frac{y_{k+1}}{y_{k+1}^{(1)}} - \frac{y_k}{y_k^{(1)}}\right) = R_k. \tag{3.126}$$

Multiplying by $y_{k+1}^{(1)}$ gives

$$y_{k+1}^{(1)}y_{k+2} - y_{k+2}^{(1)}y_{k+1} - q_k y_k^{(1)}y_{k+1} + q_k y_{k+1}^{(1)}y_k = R_k y_{k+1}^{(1)}, \tag{3.127}$$

or

$$(E - q_k)(y_k^{(1)}y_{k+1} - y_{k+1}^{(1)}y_k) = R_k y_{k+1}^{(1)}. \tag{3.128}$$

Consequently, knowledge of one solution, $y_k^{(1)}$, to the homogeneous equation allows us to factorize the left-hand side of the inhomogeneous equation (3.115) when the equation is multiplied by $y_{k+1}^{(1)}$. Note that since the coefficients of the operators that appear in equation (3.128) are functions of k, the operators do not commute; therefore, the order of the operators, as given on the left-hand side of equation (3.128), cannot be changed.

It should also be pointed out that in going from equation (3.115) to equation (3.120), we have reduced the order of the difference equation to be solved by one. In other words, equation (3.115) is a second-order equation for y_k, while equation (3.120) is a first-order equation for u_k. This procedure can be applied to higher-order equations. Thus, if any particular solution of the homogeneous equation

$$y_{k+n} + a_1(k)y_{k+n-1} + \cdots + a_n(k)y_k = 0 \tag{3.129}$$

is known, then the order of the inhomogeneous equation

$$y_{k+n} + a_1(k)y_{k+n-1} + \cdots + a_n(k)y_k = R_k \tag{3.130}$$

can be reduced by unity.

Furthermore, suppose that equation (3.131) can be operationally factorized in the form

$$[\alpha_1(k)E + \beta_1(k)][\alpha_2(k)E + \beta_2(k)] \cdots [\alpha_n(k)E + \beta_n(k)]y_k = R_n, \tag{3.131}$$

where the $\alpha_i(k)$, $\beta_i(k)$ are known functions of k. The general solution can be determined by solving, in turn, n linear difference equations each in one dependent variable. This may be seen by letting

$$z_1(k) = [\alpha_2(k)E + \beta_2(k)] \cdots [\alpha_n(k)E + \beta_n(k)]y_k, \tag{3.132}$$

and solving the equation

$$[\alpha_1(k)E + \beta_1(k)]z_1(k) = R_k \tag{3.133}$$

for $z_1(k)$. Now let

$$[\alpha_3(k)E + \beta_3(k)] \cdots [\alpha_n(k)E + \beta_n(k)]y_k = z_2(k), \tag{3.134}$$

and solve the equation

$$[\alpha_2(k)E + \beta_2(k)]z_2(k) = z_1(k). \tag{3.135}$$

Continuing this procedure n times, we will arrive at an expression for y_k with the proper number of arbitrary constants.

We end this section with a method for obtaining a solution of the second-order, homogeneous difference equation when y_0 and y_1 are prescribed. Consider the equation

$$H_k y_{k+1} + G_k y_k + F_k y_{k-1} = 0, \qquad n = 1, 2, 3, \ldots, \tag{3.136}$$

where H_k, G_k, and F_k are given functions of k, with $H_k \neq 0$. We wish to determine the solution when y_0 and y_1 are known. Equation (3.136) can be reduced to a first-order equation by dividing by $H_k y_k$, i.e.,

$$\frac{y_{k+1}}{y_k} + \frac{G_k}{H_k} + \frac{F_k/H_k}{y_k/y_{k-1}} = 0. \tag{3.137}$$

This is a nonlinear first-order equation for y_{k+1}/y_k; it can be solved by repeated iteration. The solution can be expressed as

$$y_{k+1}/y_k = p_k, \tag{3.138}$$

where the function p_k is defined by means of the following continued fraction:

$$-p_k = G_k/H_k + \cfrac{F_k/G_k}{G_{k+1}/H_{k+1} + \cfrac{F_{k-1}/H_{k-1}}{G_{k-2}/H_{k-2} + \cdots \cfrac{\cdots}{\cfrac{F_1/H_1}{-y_1/y_0}}}}. \tag{3.139}$$

Now when $k = 0$, $p_0 = y_1/y_0$. Therefore, p_k as evaluated from equation (3.139) determines the ratio of successive y_k as seen from equation (3.138). Since

$$y_{k+1} = \frac{y_{k+1}}{y_k}\frac{y_k}{y_{k-1}}\cdots\frac{y_1}{y_0}y_0, \tag{3.140}$$

we have

$$y_{k+1} = y_0 \prod_{i=0}^{k} p_i. \tag{3.141}$$

The evaluation of the product can be made easier by using the fact that p_k satisfies the equation

$$p_k = -\frac{G_k}{H_k} - \frac{F_k}{H_k p_{k-1}}. \tag{3.142}$$

This result follows from equations (3.138) and (3.139).

If y_0 and y_1 are both nonzero, then y_{k+1} is given by equation (3.141). Also, if $y_0 = y_1 = 0$, then the solution is $y_k = 0$ for all k. When $y_0 = 0$ and $y_1 \neq 0$, equation (3.140) can be rewritten so as not to depend on y_0, i.e.,

$$y_{k+1} = \frac{y_{k+1}}{y_k}\frac{y_k}{y_{k-1}}\cdots\frac{y_2}{y_1}y_1 \tag{3.143}$$

and

$$y_{k+1} = y_1 \prod_{i=1}^{k} p_i. \tag{3.144}$$

Likewise, when $y_0 \neq 0$ and $y_1 = 0$, we have

$$y_{k+1} = y_2 \prod_{i=2}^{k} p_i = -\frac{F_1}{H_1} y_0 \prod_{i=2}^{k} p_i, \tag{3.145}$$

where y_2 has been expressed in terms of y_0 by means of equation (3.136).

3.5.1. Example A

Assume that we know that one solution of the equation

$$y_{k+2} - k(k+1)y_k = 0 \tag{3.146}$$

is $y_k^{(1)} = (k-1)!$. We wish to find a second linearly independent solution. From equation (3.146), we see that

$$q_k = -k(k+1). \tag{3.147}$$

Substitution of this into equations (3.111) and (3.113) gives

$$y_k^{(2)} = (-1)^k (k-1)!. \tag{3.148}$$

Therefore, the general solution to equation (3.146) is

$$y_k = [c_1 + c_2(-1)^k](k-1)!, \tag{3.149}$$

where c_1 and c_2 are arbitrary constants. Note that the Casoratian is

$$C(k) = (-1)^{k+1}[(k-1)!]^2(2k), \tag{3.150}$$

thus showing that $y_k^{(1)}$ and $y_k^{(2)}$ are linearly independent.

3.5.2. Example B

It is straightforward to show that $y_k^{(1)} = k - 1$ is a solution to the homogeneous equation

$$y_{k+2} - \frac{2k+1}{k} y_{k+1} + \frac{k}{k-1} y_k = 0. \tag{3.151}$$

We will now use this to determine the general solution to the inhomogeneous equation

$$y_{k+2} - \frac{2k+1}{k} y_{k+1} + \frac{k}{k-1} y_k = k(k+1). \tag{3.152}$$

With the identification

$$y_k = y_k^{(1)} u_k, \qquad q_k = \frac{k}{k-1}, \qquad R_k = k(k+1), \tag{3.153}$$

equation (3.120) becomes

$$(k+1)\Delta u_{k+1} - k\Delta u_k = k(k+1). \tag{3.154}$$

This equation has the solution

$$\Delta u_k = A/k + \tfrac{1}{3}(k^2 - 1), \tag{3.155}$$

where A is an arbitrary constant. If we define

$$\phi(k) = \sum_{i=1}^{k-1} \frac{1}{i}, \tag{3.156}$$

then equation (3.155) has the solution

$$u_k = A\phi(k) + B + \tfrac{1}{18}k(k+1)(2k-5), \tag{3.157}$$

where B is a second arbitrary constant. Therefore, the general solution of (3.152) is

$$\begin{aligned} y_k &= y_k^{(1)} u_k = (k-1)u_k \\ &= A\phi(k)(k-1) + B(k-1) + \tfrac{1}{18}(k-1)k(k+1)(2k-5). \end{aligned} \tag{3.158}$$

3.5.3. Example C

The second-order equation

$$y_{k+2} - y_{k+1} - k^2 y_k = 0 \tag{3.159}$$

can be written in the following factored form:

$$(E + k)(E - k)y_k = 0. \tag{3.160}$$

Note that

$$(E - k)(E + k)y_k = y_{k+2} + y_{k+1} - k^2 y_k, \tag{3.161}$$

which shows that the order of the factors is important.

To solve equation (3.160), let

$$z_1(k) = (E - k)y_k; \tag{3.162}$$

therefore, equation (3.160) becomes

$$(E + k)z_1(k) = 0, \tag{3.163}$$

which has the solution

$$z_1(k) = A(-1)^k (k - 1)!, \tag{3.164}$$

where A is an arbitrary function. We now have

$$(E - k)y_k = A(-1)^k (k - 1)!, \tag{3.165}$$

or

$$y_{k+1} - k y_k = A(-1)^k (k - 1)!. \tag{3.166}$$

The solution to this latter equation is

$$y_k = (k - 1)! \left(\sum_{i=1}^{k-1} \frac{A(-1)^i (i - 1)!}{i!} + B \right) \cdot \tag{3.167}$$

Therefore, the general solution to equation (3.159) is

$$y_k = \left[A \sum_{i=1}^{k-1} \frac{(-1)^i}{i} + B \right] (k - 1)!. \tag{3.168}$$

3.6. STURM–LIOUVILLE DIFFERENCE EQUATIONS

Consider the following linear, homogeneous, second-order difference equation:

$$A_k y_{k+1} + B_k y_k + C_k y_{k-1} = 0, \tag{3.169}$$

where A_k, B_k, and C_k are given functions of k. This equation can always be written in the form

$$\Delta(p_{k-1}\Delta y_{k-1}) + s_k y_k = 0 \tag{3.170}$$

by the proper choice of the functions p_k and s_k. To show this, expand equation (3.170) to give

$$p_k y_{k+1} - (p_k + p_{k-1} - s_k)y_k + p_{k-1}y_{k-1} = 0. \tag{3.171}$$

Comparison of equations (3.169) and (3.171) gives

$$p_k = A_k, \qquad p_k + p_{k-1} - s_k = -B_k, \qquad p_{k-1} = C_k. \tag{3.172}$$

Taking the ratio of the first and third expressions gives

$$p_k = \frac{A_k}{C_k} p_{k-1}. \tag{3.173}$$

Solving the second expression for s_k and using the fact that

$$p_k/A_k = 1 \tag{3.174}$$

gives

$$s_k = p_k + p_{k-1} + B_k = p_k + p_{k-1} + (B_k/A_k)p_k. \tag{3.175}$$

Note that since A_k, B_k, and C_k are given, equation (3.173) can be used to determine p_k. It is specified uniquely up to an arbitrary multiplicative constant. The definition of s_k, in equation (3.175), implies also that s_k is only defined up to the same multiplicative constant. Since p_k and s_k appear linearly in equation (3.170), a unique expression is obtained.

Suppose s_k can be expressed in the form

$$s_k = q_k + \lambda r_k, \tag{3.176}$$

where λ is independent of k. Under this condition equation (3.170) becomes

$$\Delta(p_{k-1}\Delta y_{k-1}) + (q_k + \lambda r_k)y_k = 0. \tag{3.177}$$

A Sturm–Liouville difference equation is one that can be written as in equation (3.177) where $1 \le k \le N - 1$. Associated with this difference equation are the following boundary conditions:

$$a_0 y_0 + a_1 y_1 = 0, \qquad a_N y_N + a_{N+1} y_{N+1} = 0, \tag{3.178}$$

where a_0, a_1, a_N, and a_{N+1} are given constants.

Our task is to determine nontrivial solutions of equation (3.177) subject to the constraints of equation (3.178); this is the Sturm–Liouville boundary-value problem. Equations (3.177) and (3.178) constitute a Sturm–Liouville system.

In general, nontrivial solutions to the Sturm–Liouville boundary-value problem exist only if λ takes on one of a set of characteristic values or eigenvalues. The solutions corresponding to these characteristic values or eigenvalues are called characteristic functions or eigenfunctions. If $\{\lambda_i\}$ denotes the set of eigenvalues, then we will represent the corresponding eigenfunctions by $\{\phi_{i,k}\}$. Thus, for example, if λ_m and λ_n are two different eigenvalues, the corresponding eigenfunctions are, respectively, $\phi_{m,k}$ and $\phi_{n,k}$.

We now state and prove a number of results concerning the eigenvalues and eigenfunctions of the Sturm–Liouville problem. We assume, without loss of generality, that p_k, q_k, and r_k are real, and that $r_k > 0$.

Result 1: Let λ_m and λ_n be two different eigenvalues and let $\phi_{m,k}$ and $\phi_{n,k}$ be the corresponding eigenfunctions; then the following orthogonality condition is satisfied:

$$\sum_{k=1}^{N} r_k \phi_{m,k} \phi_{n,k} = 0. \tag{3.179}$$

By definition, we have

$$\Delta(p_{k-1}\Delta\phi_{m,k-1}) + (q_k + \lambda_m r_k)\phi_{m,k} = 0, \tag{3.180}$$

$$\Delta(p_{k-1}\Delta\phi_{n,k-1}) + (q_k + \lambda_n r_k)\phi_{n,k} = 0. \tag{3.181}$$

If we multiply these equations, respectively, by $\phi_{n,k}$ and $\phi_{m,k}$, subtract one from the other, and sum from $k = 1$ to N, then the following expression is obtained:

$$(\lambda_m - \lambda_n) \sum_{k=1}^{N} r_k \phi_{m,k} \phi_{n,k} = \sum_{k=1}^{N} \phi_{m,k} \Delta(p_{k-1}\Delta\phi_{n,k-1}) - \sum_{k=1}^{N} \phi_{n,k} \Delta(p_{k-1}\Delta\phi_{n,k-1}). \tag{3.182}$$

Applying summation by parts to the right-hand side of equation (3.182) gives

$$(\lambda_m - \lambda_n) \sum_{k=1}^{N} r_k \phi_{m,k} \phi_{n,k} = p_{k-1}(\phi_{m,k} \Delta\phi_{n,k-1} - \phi_{n,k} \Delta\phi_{m,k-1})\Big|_1^{N+1}$$

$$= p_N(\phi_{m,N}\phi_{n,N+1} - \phi_{n,N}\phi_{m,N+1}) \tag{3.183}$$

$$- p_0(\phi_{m,0}\phi_{n,1} - \phi_{n,0}\phi_{m,1}).$$

The boundary conditions, given by equation (3.178), yield the following results:

$$\begin{aligned} a_0\phi_{m,0} + a_1\phi_{m,1} &= 0, \\ a_0\phi_{n,0} + a_1\phi_{n,1} &= 0, \\ a_N\phi_{m,N} + a_{N+1}\phi_{m,N+1} &= 0, \\ a_N\phi_{n,N} + a_{N+1}\phi_{n,N+1} &= 0. \end{aligned} \tag{3.184}$$

Now, the constants a_0, a_1, a_N, and a_{N+1} can be eliminated from these equations to give the two relations

$$\begin{aligned} \phi_{m,0}\phi_{n,1} - \phi_{n,0}\phi_{m,1} &= 0, \\ \phi_{m,N}\phi_{n,N+1} - \phi_{n,N}\phi_{m,N+1} &= 0. \end{aligned} \tag{3.185}$$

Comparison of the right-hand side of equation (3.183) and equations (3.185) allows the conclusion

$$(\lambda_m - \lambda_n) \sum_{k=1}^{N} r_k \phi_{m,k} \phi_{n,k} = 0. \tag{3.186}$$

Since $\lambda_m \neq \lambda_n$, we must have

$$\sum_{k=1}^{N} r_k \phi_{m,k} \phi_{n,k} = 0. \tag{3.187}$$

The condition expressed by equation (3.187) states that with respect to the density function r_k, the set of functions $\{\phi_{n,k}\}$ is orthogonal. Note that the expression

$$\sum_{k=1}^{N} r_k \phi_{n,k}^2 \tag{3.188}$$

is, in general, nonzero. This follows from the fact that both r_k and $\phi_{n,k}^2$ are non-negative quantities. Therefore, we can always normalize the eigenfunctions so that they satisfy the relation

$$\sum_{k=1}^{N} r_k \phi_{m,k} \phi_{n,k} = \delta_{m,n}, \tag{3.189}$$

where $\delta_{m,n}$ is zero for $m \neq n$ and one for $m = n$. The set $\{\sqrt{r_k}\, \phi_{n,k}\}$ is called an orthonormal set.

Result 2: If p_k, q_k, and r_k are real functions, then the eigenvalues are real.

Assume that the eigenvalue λ_m and the corresponding eigenfunction $\phi_{m,k}$ are both complex. Then we have

$$\Delta(p_{k-1}\Delta\phi_{m,k-1}) + (q_k + \lambda_m r_k)\phi_{m,k} = 0, \tag{3.190}$$

and

$$\Delta(p_{k-1}\Delta\phi_{m,k-1}{}^*) + (q_k + \lambda_m{}^* r_k)\phi_{m,k}{}^* = 0, \tag{3.191}$$

where the complex conjugate operation is indicated by a star. If equations (3.190) and (3.191) are multiplied, respectively, by $\phi_{m,k}{}^*$ and $\phi_{m,k}{}^*$; if they are then subtracted and the resulting expression summed from $k = 1$ to N; and if the boundary conditions are used, then we obtain

$$(\lambda_m - \lambda_m{}^*) \sum_{k=1}^{N} r_k |\phi_{m,k}|^2 = 0. \tag{3.192}$$

The sum in equation (3.192) is positive since $r_k > 0$ and $|\phi_{m,k}|^2 \geq 0$. Therefore, we must have

$$\lambda_m - \lambda_m{}^* = 0, \tag{3.193}$$

which implies that λ_m is real.

Result 3: Let $\phi_{m,k}$ and $\phi_{n,k}$ be real and not identically zero; let r_k be real and positive; let $\phi_{m,k}$ and $\phi_{n,k}$ be orthogonal relative to the density function r_k; then $\phi_{m,k}$ and $\phi_{n,k}$ are linearly independent over the range $k = 1, 2, \ldots, N$.

Assume to the contrary that $\phi_{m,k}$ and $\phi_{n,k}$ are linearly dependent. Then there exist constants c_1 and c_2, both not zero, such that

$$c_1\phi_{m,k} + c_2\phi_{n,k} = 0. \tag{3.194}$$

Multiplying this expression by $r_k \phi_{m,k}$ and summing gives

$$c_1 \sum_{k=1}^{N} r_k \phi_{m,k}{}^2 + c_2 \sum_{k=1}^{N} r_k \phi_{m,k} \phi_{n,k} = 0, \tag{3.195}$$

which is equal to

$$c_1 \cdot 1 + c_2 \delta_{m,n} = 0 \tag{3.196}$$

if the eigenfunctions have the normalization given in equation (3.189). For $m \neq n$, equation (3.196) implies that $c_1 = 0$, which further implies from equation (3.194) that $c_2 = 0$. Consequently, we conclude that $\phi_{m,k}$ and $\phi_{n,k}$ are linearly independent.

Thus far, we have assumed, for $n \neq m$, that $\lambda_n \neq \lambda_m$. This may not always be the situation; that is, we may have $n \neq m$ but $\lambda_n = \lambda_m$. Such a case is labeled degenerate. If $\lambda_n = \lambda_m$, then the sum in equation (3.186) need not vanish. This means that linearly independent eigenfunctions corresponding to the same eigenvalue are not necessarily orthogonal. While the eigenfunctions in this degenerate case may not be orthogonal, they can always be made orthogonal. We now illustrate how this can be done.

Let the eigenfunctions $\phi_{n\,1,k}$ and $\phi_{n\,2,k}$ have, respectively, eigenvalues $\lambda_{n\,1}$ and $\lambda_{n\,2}$. Assume that $\lambda_{n\,1} = \lambda_{n\,2}$. Note that while $\phi_{n\,1,k}$ and $\phi_{n\,2,k}$ may be linearly independent, in general, we do not expect them to be orthogonal. To proceed, it is always possible to normalize $\phi_{n\,1,k}$ such that

$$\sum_{k=1}^{N} r_k \phi_{n\,1,k}{}^2 = 1. \tag{3.197}$$

Now define the function

$$\bar{\phi}_{n\,1,k} = \alpha\phi_{n\,1,k} + \beta\phi_{n\,2,k}, \tag{3.198}$$

where α and β are constants. Make the following two requirements:

$$\sum_{k=1}^{N} r_k \bar{\phi}_{n\,1,k}{}^2 = 1 \tag{3.199}$$

and

$$\sum_{k=1}^{N} r_k \bar{\phi}_{n\,1,k} \phi_{n\,1,k} = 0; \tag{3.200}$$

that is, $\bar{\phi}_{n\,1,k}$ is normalized and it is orthogonal to $\phi_{n\,1,k}$. Equations (3.199) and (3.200) give us two equations that can be solved for α and β. Note that both $\phi_{n1,k}$ and $\bar{\phi}_{n1,k}$ have the eigenvalue $\lambda_{n1} = \lambda_{n2}$, but they are orthogonal to each other. This procedure can be generalized to higher-order degeneracies.

It can be shown that the general Sturm–Liouville boundary-value problem, as defined by equations (3.177) and (3.178), gives rise to N eigenvalues and to N corresponding orthogonal eigenfunctions. (For the case where degeneracies occur, the eigenfunctions can always be made orthogonal by the procedure given above.) Both the eigenvalues and eigenfunctions are real. Further, the N eigenfunctions are linearly independent over the interval from $k = 1$ to N. A major consequence of this last fact is that the eigenfunctions are complete in the sense that any function f_k, defined for $k = 1$ to N, can be written as a linear combination of the N eigenfunctions $\{\phi_{n,k}\}$. Therefore, there exist constants $\{c_n\}$ such that

$$f_k = \sum_{n=1}^{N} c_n \phi_{n,k}, \qquad k = 1, 2, \ldots, N. \tag{3.201}$$

To determine the constants $\{c_n\}$, we make use of the orthogonality of the eigenfunctions $\{\phi_{n,k}\}$. We assume that they satisfy equation (3.189). If we multiply both sides of equation (3.201) by $r_k \phi_{m,k}$ and sum from $k = 1$ to N, we obtain

$$\begin{aligned} \sum_{k=1}^{N} r_k \phi_{m,k} f_k &= \sum_{n=1}^{N} c_n \left(\sum_{k=1}^{N} r_k \phi_{m,k} \phi_{n,k} \right) \\ &= \sum_{n=1}^{N} c_n \delta_{m,n} = c_m. \end{aligned} \tag{3.202}$$

3.6.1. Example A

Consider the following Sturm–Liouville system:

$$\Delta^2 y_{k-1} + \lambda y_k = 0, \tag{3.203}$$

$$y_0 = 0, \qquad y_{N+1} = 0. \tag{3.204}$$

Comparison with equations (3.177) and (3.178) shows that $p_{k-1} = 1$, $q_k = 0$, $r_k = 1$, $a_0 = 1$, $a_1 = 0$, $a_N = 0$, and $a_{N+1} = 1$. If we let

$$2 - \lambda = 2 \cos \theta, \tag{3.205}$$

then equation (3.203) becomes

$$y_{k+1} - (2\cos\theta)y_k + y_{k-1} = 0, \tag{3.206}$$

which has the solution

$$y_k = c_1 \cos k\theta + c_2 \sin k\theta, \tag{3.207}$$

where c_1 and c_2 are arbitrary constants. The first boundary conditions $y_0 = 0$ gives $c_1 = 0$. The second boundary condition $y_{N+1} = 0$ gives

$$\sin(N+1)\theta = 0 \tag{3.208}$$

or

$$(N+1)\theta = n\pi, \qquad n = 1, 2, 3, \ldots. \tag{3.209}$$

Thus, the eigenvalues λ_n are, from equation (3.205), given by the expression

$$\begin{aligned}\lambda_n &= 2\left[1 - \cos\left[\frac{n\pi}{N+1}\right)\right] \\ &= 4\sin^2\left(\frac{n\pi}{2(N+1)}\right), \qquad n = 1, 2, 3, \ldots.\end{aligned} \tag{3.210}$$

Note that there are only N distinct values of n, i.e., $n = 1, 2, \ldots. N$, since after $n = N$ the values of the eigenvalues repeat themselves. The N eigenfunctions associated with these eigenvalues can be determined from equations (3.207), (3.209), and (3.210); they are

$$\phi_{n,k} = \sin\left(\frac{kn\pi}{N+1}\right), \qquad n = 1, 2, \ldots, N. \tag{3.211}$$

3.6.2. Example B

Let the function f_k be defined for $k = 1$ to N. Determine the coefficients $\{c_m\}$ when f_k is expanded in terms of the eigenfunctions of equation (3.211).

We have from equation (3.201)

$$f_k = \sum_{m=1}^{N} c_m \sin\left(\frac{m\pi k}{N+1}\right), \tag{3.212}$$

where c_m is determined by the relation

$$c_m = \sum_{k=1}^{N} f_k \sin\left(\frac{m\pi k}{N+1}\right) \Big/ \sum_{k=1}^{N} \sin^2\left(\frac{m\pi k}{N+1}\right). \tag{3.213}$$

Now it can be shown that

$$\sum_{k=1}^{N} \sin^2\left(\frac{m\pi k}{N+1}\right) = \frac{N+1}{2}. \tag{3.214}$$

Therefore, the coefficients $\{c_m\}$ are

$$c_m = \frac{2}{N+1} \sum_{k=1}^{N} f_k \sin\left(\frac{m\pi k}{N+1}\right), \qquad m = 1, 2, \ldots, N. \tag{3.215}$$

3.6.3. Example C

Let us determine the mth coefficient in the representation given by equation (3.212) for the function

$$f_k = \delta_{k,r}, \tag{3.216}$$

where r is a given integer such that $1 \le r \le N$. Substitution of equation (3.216) into equation (3.215) gives

$$c_m = \frac{2}{N+1} \sum_{k=1}^{N} \delta_{k,r} \sin\left(\frac{m\pi k}{N+1}\right) = \frac{2}{N+1} \sin\left(\frac{m\pi r}{N+1}\right). \tag{3.217}$$

Therefore, the function $f_k = \delta_{k,r}$ has the representation

$$f_k = \frac{2}{n+1} \sum_{m=1}^{N} \sin\left(\frac{m\pi r}{N+1}\right) \sin\left(\frac{m\pi k}{N+1}\right). \tag{3.218}$$

4
LINEAR DIFFERENCE EQUATIONS WITH CONSTANT COEFFICIENTS

4.1. INTRODUCTION

This chapter is concerned with the nth-order linear difference equation with constant coefficients,

$$y_{k+n} + a_1 y_{k+n-1} + a_2 y_{k+n-2} + \cdots + a_n y_k = R_k, \tag{4.1}$$

where the a_i are a given set of n constants, with $a_n \neq 0$, and R_k is a given function of k. If $R_k = 0$, then equation (4.1) is homogeneous:

$$y_{k+n} + a_1 y_{k+n-1} + a_2 y_{k+n-2} + \cdots + a_n y_k = 0; \tag{4.2}$$

for $R_k \neq 0$, equation (4.1) is inhomogeneous.

We know from Chapter 3 that the homogenous equation (4.2) has a fundamental set of solutions which consists of n linearly independent functions $y_k^{(i)}$, $i = 1, 2, \ldots, n$, and that the general solution is the linear combination

$$y_k^{(H)} = c_1 y_k^{(1)} + c_2 y_k^{(2)} + \cdots + c_n y_k^{(n)}, \tag{4.3}$$

where the c_i are n arbitrary constants. Likewise, the solution to the inhomogeneous equation (4.1) consists of a sum of the homogeneous solution $y_k^{(H)}$ and a particular solution to equation (4.1),

$$y_k = y_k^{(H)} + y_k^{(P)}. \tag{4.4}$$

The purpose of the chapter is to provide techniques for obtaining solutions to the homogenous equation (4.2) and the inhomogeneous equation (4.1).

4.2. HOMOGENEOUS EQUATIONS

Using the shift operator E, we can write equation (4.2)

$$f(E) y_k = 0, \tag{4.5}$$

where $f(E)$ is the operator function

$$f(E) = E^n + a_1 E^{n-1} + a_2 E^{n-2} + \cdots + a_{n-1} E + a_n. \tag{4.6}$$

Definition: The characteristic equation associated with equation (4.2) or (4.5) is

$$f(r) = r^n + a_1 r^{n-1} + a_2 r^{n-2} + \cdots + a_{n-1} r + a_n = 0. \tag{4.7}$$

Note that $f(r)$ is an nth-order polynomial and thus has n roots $\{r_i\}$, where $i = 1, 2, \ldots, n$. The function $f(r)$ can also be written in the factored form

$$f(r) = \prod_{i=1}^{n} (r - r_i) = (r - r_1)(r - r_2) \cdots (r - r_n) = 0. \tag{4.8}$$

Theorem 4.1: Let r_i be any solution to the characteristic equation (4.7); then

$$y_k = r_i{}^k \tag{4.9}$$

is a solution to the homogeneous equation (4.2).

Proof: Substituting equation (4.9) into equation (4.2) gives

$$\begin{aligned} r_i{}^{k+n} &+ a_1 r_i{}^{k+n-1} + \cdots + a_{n-1} r_i{}^{k+1} + a_n r_i{}^k \\ &= r_i{}^k (r_i{}^n + a_1 r_i{}^{n-1} + \cdots + a_n) \\ &= r_i{}^k f(r_i) = 0. \end{aligned} \tag{4.10}$$

Hence, $y_k = r_i{}^k$ is a solution to equation (4.2).

Theorem 4.2: Assume the n roots of the characteristic equations are distinct; then a fundamental set of solutions is

$$y_k{}^{(i)} = r_i{}^k, \qquad i = 1, 2, \ldots, n. \tag{4.11}$$

An immediate consequence of this theorem, for this particular case, is that the general solution to the homogeneous equation (4.2) is

$$y_k = c_1 y_k{}^{(1)} + c_2 y_k{}^{(2)} + \cdots + c_n y_k{}^{(n)}, \tag{4.12}$$

where the n constants c_i are arbitrary.

Proof: From Theorem 4.1, we know that each $y_k{}^{(i)} = r_i{}^k$ is a solution to equation (4.2). To show that the n functions of equation (4.11) form a fundamental set of solutions, we must prove that the Casoratian of these n functions is nonzero. Therefore,

$$C(k) = \begin{vmatrix} r_1{}^k & r_2{}^k & \cdots & r_n{}^k \\ r_1{}^{k+1} & r_2{}^{k+1} & \cdots & r_1{}^{k+1} \\ \cdot & \cdot & & \cdot \\ \cdot & \cdot & & \cdot \\ \cdot & \cdot & & \cdot \\ r_1{}^{k+n-1} & r_2{}^{k+n-1} & \cdots & r_n{}^{k+n-1} \end{vmatrix}$$

$$= \left(\prod_{i=1}^{n} r_i\right)^k \begin{vmatrix} 1 & 1 & \cdots & 1 \\ r_1 & r_2 & \cdots & r_n \\ \cdot & \cdot & \cdot & \\ \cdot & \cdot & \cdot & \\ \cdot & \cdot & \cdot & \\ r_1{}^{n-1} & r_2{}^{n-1} & \cdots & r_n{}^{n-1} \end{vmatrix} \tag{4.13}$$

$$= \left(\prod_{i=1}^{n} r_i\right)^k \prod_{l>m} (r_l - r_m).$$

(The last line follows from the fact that the determinant in the second line is a Vandermonde determinant.) Now, all the r_i are nonzero because of the condition $a_n \neq 0$. Furthermore, since the roots are distinct, we have $r_l - r_m \neq 0$ for $l = m$. Therefore, we conclude that $C(k)$ is never equal to zero and, consequently, the functions $y_k{}^{(i)} = r_i{}^k$, $i = 1, 2, \ldots, n$, are linearly independent and form a fundamental set of solutions.

Let us now consider the case where one or more of the roots of the characteristic equation are repeated. Assume the root r_1 has multiplicity m_1, the root r_2 has multiplicity $m_2, \ldots,$ and root r_l has multiplicity m_l, such that

$$m_1 + m_2 + \cdots + m_l = n. \tag{4.14}$$

Thus, equation (4.2) and the characteristic equation have the respective forms

$$(E - r_1)^{m_1}(E - r_2)^{m_2} \cdots (E - r_l)^{m_l} y_k = 0, \tag{4.15a}$$

$$(r - r_1)^{m_1}(r - r_2)^{m_2} \cdots (r - r_l)^{m_l} = 0. \tag{4.15b}$$

Note that to study this case, we need only investigate the individual equations

$$(E - r_i)^{m_i} y_k = 0, \qquad i = 1, 2, \ldots, l, \tag{4.16}$$

since the solutions of equation (4.16) are also solutions of equation (4.15a). To proceed, we need several results.

First, consider

$$\begin{aligned}(E - r)y_k = y_{k+1} - r y_k &= r^{k+1}\left(\frac{y_{k+1}}{r^{k+1}} - \frac{y_k}{r^k}\right)\\ &= r^{k+1}\,\Delta \frac{y_k}{r^k}.\end{aligned} \tag{4.17}$$

Therefore, it follows that

$$\begin{aligned}(E - r)^2 y_k &= r^{k+1}\,\Delta\left(\frac{1}{r^k}\, r^{k+1}\,\Delta \frac{y_k}{r^k}\right)\\ &= r^{k+2}\,\Delta^2 \frac{y_k}{r^k},\end{aligned} \tag{4.18}$$

and, for arbitrary $m = 1, 2, 3, \ldots$, we have

$$(E - r)^m y_k = r^{k+m}\,\Delta^m \frac{y_k}{r^k}. \tag{4.19}$$

Second, the equation

$$\Delta^m y_k = 0 \tag{4.20}$$

has the solution

$$y_k = A_1 + A_2 k + \cdots + A_m k^{m-1}, \tag{4.21}$$

where $A_1, A_2, \ldots, A_m$ are m arbitrary constants. Equation (4.21) is a polynomial of degree $m - 1$ in k.

Using these two results, we see that equation (4.16) has the solution

$$y_k^{(i)} = (A_1 + A_2 k + \cdots + A_{m_i} k^{m_i - 1}) r_i^k, \tag{4.22}$$

for each $i = 1, 2, \ldots, l$. Note that the solution contains m_i arbitrary constants. since equation (4.16) is an m_i th-order difference equation, this is a general solution.

In a manner similar to the calculations involving equation (4.11), we can show that the set of functions $\{r_i^k, kr_i^k, \ldots, k^{m_i-1}r_i^k\}$ are linearly independent and, thus, form a fundamental set of solutions for equation (4.16). (A quick way to see this is to note that the set of powers $\{1, k, k^2, \ldots, k^m\}$ are linearly independent over any interval $k_1 \leq k \leq k_2$. Multiplication of each power by the same function of k gives a new set of functions which are also linearly independent.) Therefore, the general solution to equation (4.15a) is

$$y_k = y_k^{(1)} + y_k^{(2)} + \cdots + y_k^{(l)}, \tag{4.23}$$

where the $y_k^{(i)}$ are given by equation (4.22).

The above results are summarized in the following theorem.

Theorem 4.3: Let

$$y_{k+n} + a_1 y_{k+n-1} + \cdots + a_{n-1} y_{k+1} + a_n y_k = 0 \tag{4.24}$$

be an nth-order linear difference equation with given constant coefficients $a_1, a_2, \ldots, a_n$ and having $a_n \neq 0$. Let the roots of the characteristic equation

$$r^n + a_1 r^{n-1} + \cdots + a_{n-1} r + a_n = 0 \tag{4.25}$$

be r_i with multiplicity m_i, $i = 1, 2, \ldots, l$, where

$$m_1 + m_2 + \cdots + m_l = n. \tag{4.26}$$

Then, the general solution of equation (4.24) is

$$\begin{aligned} y_k = {} & r_1^k (A_1^{(1)} + A_2^{(1)} k + \cdots + A_{m_1}^{(1)} k^{m_1 - 1}) \\ & + r_2^k (A_1^{(2)} + A_2^{(2)} k + \cdots + A_{m_2}^{(2)} k^{m_2 - 1}) \\ & + \cdots \\ & + r_{m_l}^k (A_1^{(l)} + A_2^{(l)} k + \cdots + A_{m_l}^{(l)} k^{m_l - 1}). \end{aligned} \tag{4.27}$$

4.2.1. Example A

The equation

$$y_{k+2} - 7y_{k+1} + 6y_k = 0 \tag{4.28}$$

can be written

$$(E^2 - 7E + 6)y_k = (E - 6)(E - 1)y_k = 0. \tag{4.29}$$

Therefore, the characteristic equation is

$$f(r) = r^2 - 7r + 6 = (r - 6)(r - 1) = 0 \tag{4.30}$$

and has the roots $r_1 = 6$, $r_2 = 1$. The two fundamental solutions are

$$y_k^{(1)} = 6^k, \qquad y_k^{(2)} = 1. \tag{4.31}$$

Therefore, the general solution is

$$y_k = c_1 6^k + c_2, \tag{4.32}$$

where c_1 and c_2 are arbitrary constants.

4.2.2. Example B

The second-order equation

$$y_{k+2} - y_k = 0 \tag{4.33a}$$

has the characteristic equation

$$r^2 - 1 = (r - 1)(r + 1) = 0, \tag{4.33b}$$

with roots $r_1 = 1$, $r_2 = -1$. The two fundamental solutions are

$$y_k^{(1)} = 1, \qquad y_k^{(2)} = (-1)^k, \tag{4.34}$$

and the general solution is

$$y_k = c_1 + c_2(-1)^k, \tag{4.35}$$

where c_1 and c_2 are arbitrary constants.

4.2.3. Example C

The characteristic equation for

$$y_{k+2} - 6y_{k+1} + 9y_k = 0 \tag{4.36}$$

is

$$r^2 - 6r + 9 = (r - 3)^2 = 0. \tag{4.37}$$

Thus, we have the double root $r_1 = r_2 = 3$ and the following two fundamental solutions:

$$y_k^{(1)} = 3^k, \qquad y_k^{(2)} = k3^k. \tag{4.38}$$

Therefore,

$$y_k = (c_1 + c_2 k)3^k, \tag{4.39}$$

where c_1 and c_2 are arbitrary constants.

4.2.4. Example D

The Fibonacci difference equation is

$$y_{k+2} = y_{k+1} + y_k. \tag{4.40}$$

The Fibonacci sequence is that solution which has $y_0 = 0$, $y_1 = 1$. The characteristic equation is

$$r^2 - r - 1 = 0, \tag{4.41}$$

and has the two roots

$$r_1 = \frac{1+\sqrt{5}}{2}, \qquad r_2 = \frac{1-\sqrt{5}}{2}. \tag{4.42}$$

Therefore, the general solution to the Fibonacci difference equation is

$$y_k = c_1\left(\frac{1+\sqrt{5}}{2}\right)^k + c_2\left(\frac{1-\sqrt{5}}{2}\right)^k, \tag{4.43}$$

where c_1 and c_2 are arbitrary constants.

The general member of the Fibonacci sequence can be determined by solving for c_1 and c_2 from the equations

$$\begin{aligned} y_0 &= c_1 + c_2 = 0, \\ y_1 &= c_1\frac{1+\sqrt{5}}{2} + c_2\frac{1-\sqrt{5}}{2} = 1. \end{aligned} \tag{4.44}$$

The solutions are $c_1 = -c_2 = 1/\sqrt{5}$. Therefore, the general member of the Fibonacci sequence is given by

$$y_k = \frac{1}{\sqrt{5}}\left[\left(\frac{1+\sqrt{5}}{2}\right)^k - \left(\frac{1-\sqrt{5}}{2}\right)^k\right]. \tag{4.45}$$

We have $\{y_k\} = 0, 1, 1, 2, 3, 5, 8, 13, 21, 34, \ldots$.

4.2.5. Example E

The third-order difference equation

$$y_{k+3} + y_{k+2} - 8y_{k+1} - 12y_k = 0 \tag{4.46}$$

has the factored characteristic function

$$r^3 + r^2 - 8r - 12 = (r-3)(r+2)^2 = 0. \tag{4.47}$$

This last equation has the roots $r_1 = 3$, $r_2 = r_3 = -2$. The corresponding fundamental set of solutions is

$$y_k^{(1)} = 3^k, \qquad y_k^{(2)} = (-2)^k, \qquad y_k^{(3)} = k(-2)^k. \tag{4.48}$$

Thus, the general solution of equation (4.46) is

$$y_k = c_1 3^k + (c_2 + c_3 k)(-2)^k, \tag{4.49}$$

where c_1, c_2, and c_3 are arbitrary constants.

4.2.6. Example F

The equation

$$y_{k+3} + y_{k+2} - y_{k+1} - y_k = 0 \tag{4.50}$$

has the characteristic equation

$$r^3 + r^2 - r - 1 = (r-1)(r+1)^2 = 0, \tag{4.51}$$

with roots $r_1 = 1$, $r_2 = r_3 = -1$. Therefore, the fundamental solutions are

$$y_k^{(1)} = 1, \qquad y_k^{(2)} = (-1)^k, \qquad y_k^{(3)} = k(-1)^k, \tag{4.52}$$

and the general solution is

$$y_k = c_1 + (c_2 + c_3 k)(-1)^k. \tag{4.53}$$

4.2.7. Example G

The equation

$$\Delta^m y_k = 0 \tag{4.54}$$

can be expressed in terms of the shift operator *E:*

$$(E - 1)^m y_k = 0. \tag{4.55}$$

The corresponding characteristic equation is

$$(r - 1)^m = 0,$$

and has m repeated solutions, all equal to 1, i.e.,

$$r_1 = r_2 = \cdots = r_{m-1} = r_m = 1. \tag{4.56}$$

Consequently, the general solution to equation (4.54) or (4.55) is

$$y_k = c_1 + c_2 k + \cdots + c_{m-1} k^{m-2} + c_m k^{m-1}, \tag{4.57}$$

where the m constants are arbitrary.

4.2.8. Example H

Consider the following second-order difference equation:

$$y_{k+2} - 2y_{k+1} - 2y_k = 0. \tag{4.58}$$

The characteristic equation

$$r^2 - 2r - 2 = 0 \tag{4.59}$$

has the two solutions

$$r_1 = 1 + i, \qquad r_2 = 1 - i, \tag{4.60}$$

where $i = \sqrt{-1}$. These two complex conjugate roots can be put in the forms

$$r_1 = \sqrt{2}\, e^{\pi i/4} = r_2^*. \tag{4.61}$$

The corresponding complex conjugate set of fundamental solutions is

$$y_k^{(1)} = 2^{k/2} e^{\pi i k/4} = y_k^{(2)*}. \tag{4.62}$$

Using

$$e^{\pm i\theta} = \cos\theta \pm i \sin\theta, \tag{4.63}$$

and taking the appropriate linear combinations, we can form another set of fundamental solutions

$$\bar{y}_k^{(1)} = 2^{k/2} \cos(\pi k/4), \qquad \bar{y}_k^{(2)} = 2^{k/2} \sin(\pi k/4). \tag{4.64}$$

Therefore, we finally obtain the general solution to equation (4.58); it is

$$y_k = c_1 2^{k/2} \cos(\pi k/4) + c_2 2^{k/2} \sin(\pi k/4). \tag{4.65}$$

Several comments are in order at this point. First, if the linear difference equation has real coefficients, then any complex roots of the characteristic equation must occur in complex conjugate pairs. Second, the corresponding fundamental solutions can be written in either of the two equivalent forms

$$y_k^{(1)} = y_k^{(2)*} = r_1^k \tag{4.66}$$

or

$$\bar{y}_k^{(1)} = R^k \cos(k\theta), \qquad \bar{y}_k^{(2)} = R^k \sin(k\theta), \tag{4.67}$$

where the complex conjugate pair of roots are

$$\begin{aligned} r_1 &= r_2^* = a + ib = Re^{i\theta}, \\ R &= \sqrt{a^2 + b^2}, \qquad \tan\theta = b/a. \end{aligned} \tag{4.68}$$

If the complex conjugate pair of roots occur with multiplicity m, then the corresponding fundamental solutions are

$$
\begin{aligned}
y_k{}^{(1)} &= r_1{}^k &&= y_k{}^{(m+1)*}, \\
y_k{}^{(2)} &= kr_1{}^k &&= y_k{}^{(m+2)*}, \\
&\;\;\vdots && \;\;\vdots \\
y_k{}^{(m)} &= k^{m-1}r_1{}^k &&= y_k{}^{(2m)*}.
\end{aligned}
\tag{4.69}
$$

(Note that there are $2m$ roots; therefore, the above labeling of the fundamental solutions should be clear.) An equivalent form for these solutions is

$$
\begin{aligned}
\bar{y}_k{}^{(1)} &= R^k \cos(k\theta), & \bar{y}_k{}^{(m+1)} &= R^k \sin(k\theta), \\
\bar{y}_k{}^{(2)} &= kR^k \cos(k\theta), & \bar{y}_k{}^{(m+2)} &= kR^k \sin(k\theta), \\
&\;\;\vdots & &\;\;\vdots \\
\bar{y}_k{}^{(m)} &= k^{m-1}R^k \cos(k\theta), & \bar{y}_k{}^{(2m)} &= k^{m-1}R^k \sin(k\theta).
\end{aligned}
\tag{4.70}
$$

4.2.9 Example I

The equation

$$k_{k+4} - y_k = 0 \tag{4.71}$$

has the characteristic equation

$$r^4 - 1 = (r^2 + 1)(r^2 - 1) = (r + i)(r - i)(r + 1)(r - 1), \tag{4.72}$$

and roots $r_1 = -i$, $r_2 = +i$, $r_3 = -1$, $r_4 = +1$. Since $\exp(\pm i\pi/2) = \pm i$, we have the following fundamental set of solutions:

$$
\begin{aligned}
y_k{}^{(1)} &= e^{-i\pi k/2}, & y_k{}^{(2)} &= e^{i\pi k/2}, \\
y_k{}^{(3)} &= (-1)^k, & y_k{}^{(4)} &= 1.
\end{aligned}
\tag{4.73}
$$

Note that $y_k{}^{(1)}$ and $y_k{}^{(2)}$ can be written in the equivalent forms

$$\bar{y}_k{}^{(1)} = \cos(\pi k/2), \qquad \bar{y}_k{}^{(2)} = \sin(\pi k/2). \tag{4.74}$$

Therefore, the general solution to equation (4.71) is

$$y_k = c_1 \cos(\pi k/2) + c_2 \sin(\pi k/2) + c_3(-1)^k + c_4, \tag{4.75}$$

where c_1, c_2, c_3, and c_4 are arbitrary constants.

4.2.10. Example J

Now consider the fourth-order equation

$$y_{k+4} + y_k = 0. \tag{4.76}$$

Its characteristic equation is

$$r^4 + 1 = (r^2 + i)(r^2 - i) = 0. \tag{4.77}$$

From the first factor, we obtain

$$r_1 = e^{-i\pi/4}, \qquad r_2 = e^{-3i\pi/4}, \tag{4.78}$$

and from the second factor

$$r_3 = e^{i\pi/4}, \qquad r_4 = e^{3i\pi/4}. \tag{4.79}$$

Since $r_1 = r_3^*$ and $r_2 = r_4^*$, we have the following four fundamental solutions:

$$\begin{aligned} y_k^{(1)} &= \cos(\pi k/4), \qquad y_k^{(2)} = \sin(\pi k/4), \\ y_k^{(3)} &= \cos(3\pi k/4), \quad y_k^{(4)} = \sin(3\pi k/4). \end{aligned} \tag{4.80}$$

Consequently, the general solution to equation (4.71) is

$$\begin{aligned} y_k = {} & c_1 \cos(\pi k/4) + c_2 \sin(\pi k/4) \\ & + c_3 \cos(3\pi k/4) + c_4 \sin(3\pi k/4), \end{aligned} \tag{4.81}$$

where c_1, c_2, c_3, and c_4 are arbitrary constants.

4.2.11. Example K

Consider the equation

$$y_{k+1} - (2\cos\phi) y_k + y_{k-1} = 0, \qquad \cos\phi \neq 0, \tag{4.82}$$

where ϕ is a constant. The characteristic equation is

$$r^2 - (2 \cos \phi)r + 1 = 0, \tag{4.83}$$

and has the solutions

$$\begin{aligned} r_\pm &= \tfrac{1}{2}(2 \cos \phi \pm \sqrt{4 \cos^2 \phi - 4}) \\ &= \cos \phi \pm i \sin \phi = e^{\pm i\phi}. \end{aligned} \tag{4.84}$$

Therefore, a fundamental set of solutions is

$$y_k^{(1)} = \cos(\phi k), \qquad y_k^{(2)} = \sin(\phi k), \tag{4.85}$$

and the general solution is

$$y_k = c_1 \cos(\phi k) + c_2 \sin(\phi k), \tag{4.86}$$

where c_1 and c_2 are arbitrary constants.

4.2.12. Example L

Let a ninth-order difference equation have the following roots to its characteristic equation:

$$\begin{aligned} &r_1 = r_2 = r_3 = 6, \qquad r_4 = -1, \qquad r_5 = \frac{1}{2}, \\ &r_6 = r_7 = r_8{}^* = r_9{}^* = -\frac{1}{2} + \frac{\sqrt{3}}{2} i. \end{aligned} \tag{4.87}$$

The fundamental set of solutions for this equation is

$$\begin{aligned} &y_k^{(1)} = 6^k, \qquad y_k^{(2)} = k6^k, \qquad y_k^{(3)} = k^2 6^k, \\ &y_k^{(4)} = (-1)^k, \qquad y_k^{(5)} = 2^{-k}, \\ &y_k^{(6)} = \cos(2\pi k/3), \qquad y_k^{(7)} = k \cos(2\pi k/3), \\ &y_k^{(8)} = \sin(2\pi k/3), \qquad y_k^{(9)} = k \sin(2\pi k/3). \end{aligned} \tag{4.80}$$

To obtain the results given in the last two lines of equation (4.88), we used

$$-\frac{1}{2} \pm \frac{\sqrt{3}}{2} i = \cos \theta \pm i \sin \theta = e^{\pm i\theta}, \qquad \theta = \frac{2\pi}{3} \cdot \tag{4.89}$$

We conclude that this ninth-order difference equation has the general solution

$$\begin{aligned} y_k = {} & (c_1 + c_2 k + c_3 k^2) 6^k + c_4 (-1)^k + c_5 2^{-k} \\ & + (c_6 + c_7 k)\cos(2\pi k/3) + (c_8 + c_9 k)\sin(2\pi k/3), \end{aligned} \tag{4.90}$$

where c_i, $i = 1, 2, \ldots, c_9$, are arbitrary constants.

4.3. CONSTRUCTION OF A DIFFERENCE EQUATION HAVING SPECIFIED SOLUTIONS

Suppose we are given n linearly independent functions $y_k^{(i)}$, $i = 1, 2, \ldots, n$. An nth-order linear homogeneous difference equation having these functions as a fundamental system is given by the determinant equation

$$\begin{vmatrix} y_k & y_k^{(1)} & y_k^{(2)} & \cdots & y_k^{(n)} \\ y_{k+1} & y_{k+1}^{(1)} & y_{k+1}^{(2)} & \cdots & y_{k+1}^{(n)} \\ \vdots & \vdots & \vdots & & \vdots \\ y_{k+n} & y_{k+n}^{(1)} & y_{k+n}^{(2)} & \cdots & y_{k+n}^{(n)} \end{vmatrix} = 0. \tag{4.91}$$

This result follows from the fact that the determinant is zero when $y_k = y_k^{(i)}$ for $i = 1, 2, \ldots, n$.

4.3.1. Example A

A first-order difference equation having the solution

$$y_k = e^{-hk}, \qquad h = \text{constant}, \tag{4.92}$$

is

$$\begin{vmatrix} y_k & e^{-hk} \\ y_{k+1} & e^{-h(k+1)} \end{vmatrix} = e^{-hk} \begin{vmatrix} y_k & 1 \\ y_{k+1} & e^{-h} \end{vmatrix} = 0, \tag{4.93}$$

or

$$y_{k+1} - e^{-h} y_k = 0. \tag{4.94}$$

An easy calculation shows that this last equation has the general solution

$$y_k = ce^{-hk}, \tag{4.95}$$

where c is an arbitrary constant.

4.3.2. Example B

Consider the two linearly independent functions

$$y_k^{(1)} = 1, \qquad y_k^{(2)} = 3^k. \tag{4.96}$$

A second-order equation having these solutions is

$$\begin{vmatrix} y_k & 1 & 3^k \\ y_{k+1} & 1 & 3^{k+1} \\ y_{k+2} & 1 & 3^{k+2} \end{vmatrix} = 3^k \begin{vmatrix} y^k & 1 & 1 \\ y_{k+1} & 1 & 3 \\ y_{k+2} & 1 & 9 \end{vmatrix} = 0, \tag{4.97}$$

or

$$y_{k+2} - 4y_{k+1} + 3y_k = 0. \tag{4.98}$$

4.3.3. Example C

The following two functions are linearly independent:

$$y_k^{(1)} = e^{i\phi k}, \qquad y_k^{(2)} = e^{-i\phi k}, \qquad \phi = \text{constant}. \tag{4.99}$$

To determine the second-order difference equation that has these two functions as solutions, form the determinant

$$\begin{aligned} \begin{vmatrix} y_k & 1 & 1 \\ y_{k+1} & e^{i\phi} & e^{-i\phi} \\ y_{k+2} & e^{2i\phi} & e^{-2i\phi} \end{vmatrix} &= (-2i \sin \phi) \left(y_{k+2} - \frac{\sin(2\phi)}{\sin \phi} y_{k+1} + y_k \right) \\ &= (-2i \sin \phi)[y_{k+2} - (2 \cos \phi) y_{k+1} + y_k]. \end{aligned} \tag{4.100}$$

Setting the determinant equal to zero gives

$$y_{k+2} - (2\cos\phi)y_{k+1} + y_k = 0, \tag{4.101}$$

which is the desired equation.

4.3.4 Example D

The second-order difference equation whose fundamental system of solutions is

$$y_k^{(1)} = k, \qquad y_k^{(2)} = e^{-hk} \tag{4.102}$$

is

$$\begin{vmatrix} y_k & k & e^{-hk} \\ y_{k+1} & k+1 & e^{-h(k+1)} \\ y_{k+2} & k+2 & e^{-h(k+2)} \end{vmatrix} = 0. \tag{4.103}$$

Expanding the determinant and simplifying the resulting expression gives

$$[k(1-e^{-h})+1]y_{k+2} - [k(1-e^{-2h})+2]y_{k+1} + e^{-h}[k(1-e^{-h})+2-e^{-h}]y_k = 0. \tag{4.104}$$

4.3.5. Example E

Let us determine the second-order difference equation having the fundamental system

$$y_k^{(1)} = k!, \qquad y_k^{(2)} = 1. \tag{4.105}$$

We have

$$\begin{vmatrix} y_k & k! & 1 \\ y_{k+1} & (k+1)! & 1 \\ y_{k+2} & (k+2)! & 1 \end{vmatrix} = k! \begin{vmatrix} y_k & 1 & 1 \\ y_{k+1} & k+1 & 1 \\ y_{k+2} & (k+2)(k+1) & 1 \end{vmatrix} = 0, \tag{4.106}$$

which gives

$$ky_{k+2} - (k^2+3k+1)y_{k+1} + (k+1)^2 y_k = 0. \tag{4.107}$$

4.4. RELATIONSHIP BETWEEN LINEAR DIFFERENCE AND DIFFERENTIAL EQUATIONS

Consider an nth-order linear homogeneous differential equation having constant coefficients,

$$D^n y(x) + a_1 D^{n-1} y(x) + \cdots + a_n y(x) = 0, \tag{4.108}$$

where $D \equiv d/dx$ is the differentiation operator, the a_i, $i = 1, 2, \ldots, n$, are given constants, and $a_n \neq 0$. Associated with this differential equation is the following difference equation:

$$y_{k+n} + a_1 y_{k+n-1} + \cdots + a_n y_k = 0. \tag{4.109}$$

The following theorem makes clear the relationship between the solutions of the two equations.

Theorem 4.4: Let

$$y(x) = \sum_{i=1}^{l} \left(\sum_{j=0}^{n_i - 1} c_{i,j+1} x^j \right) e^{r_i x} + \sum_{j=(n_1 + \cdots + n_l)+1}^{n} c_j e^{r_j x} \tag{4.110}$$

be the general solution of equation (4.108), where $c_{i,j+1}$ and c_j are arbitrary constants; $n_i \geq 1$, $i = 1, 2, \ldots, l$, with $n_1 + n_2 + \cdots + n_l \leq n$; and where the characteristic equation

$$r^n + a_1 r^{n-1} + \cdots + a_n = 0 \tag{4.111}$$

has roots r_i with multiplicity n_i, $i = 1, 2, \ldots, l$, and the simple roots r_j.

Let y_k be the general solution of equation (4.109). Then

$$y_k = D^k y(x)|_{x=0}, \tag{4.112}$$

and

$$y_k = \sum_{i=1}^{l} \left(c_{i1} + \sum_{m=1}^{n_i - 1} \gamma_{i,m} k^m r_i^{\,k} \right) + \sum_{j=(n_1 + \cdots + n_l)+1}^{n} c_j r_j^{\,k}, \tag{4.113}$$

where the $\gamma_{i,m}$ are arbitrary constants.

Proof: If equation (4.108) is differentiated k times with respect to x and x is set equal to zero, then equation (4.109) is obtained, where y_k is given by equation (4.112). If now equation (4.110) is differentiated k times and x is set equal to zero, then the result of equation (4.113) follows.

Note that the choice of $x = 0$ in the theorem is arbitrary and is useful primarily because it is convenient in the evaluation of y_k. In general, any other value of x could be used.

4.4.1. Example A

The second-order differential equation

$$\frac{d^2y}{dx^2} - 3\frac{dy}{dx} + 2y = 0 \tag{4.114}$$

has the general solution

$$y(x) = c_1 e^x + c_2 e^{2x}, \tag{4.115}$$

where c_1 and c_2 are arbitrary constants. The difference equation associated with this differential equation is

$$y_{k+2} - 3y_{k+1} + 2y_k = 0. \tag{4.116}$$

Its general solution is

$$y_k = A + B2^k, \tag{4.117}$$

since the characteristic equation $r^2 - 3r + 2 = 0$ has roots $r_1 = 1$ and $r_2 = 2$; A and B are arbitrary constants. We now show how the result given by equation (4.117) can be obtained from equation (4.115).

Let us calculate $D^k y(x)$; it is

$$D^k y(x) = \frac{d^n}{dx^n}(c_1 e^x + c_2 e^{2x}) = c_1 e^x + c_2 2^k e^{2x}. \tag{4.118}$$

Therefore,

$$y_k = D^k y(x)|_{x=0} = c_1 + c_2 2^k, \tag{4.119}$$

which is the same as equation (4.117) except for the labeling of the arbitrary constants.

4.4.2. Example B

The differential equation

$$\frac{d^2y}{dx^2} - 2\frac{dy}{dx} + y = 0 \tag{4.120}$$

has the general solution

$$y(x) = (c_1 + c_2 x)e^x = c_1 e^x + c_2 x e^x. \tag{4.121}$$

The associated difference equation is

$$y_{k+2} - 2y_{k+1} + y_k = 0. \tag{4.122}$$

From equation (4.121) we obtain

$$\begin{aligned} D^k y(x) &= \frac{d^k}{dx^k}(c_1 e^x + c_2 x e^x) \\ &= c_1 e^x + c_2(x e^x + k e^x), \end{aligned} \tag{4.123}$$

where the expression in parentheses on the right-hand side of equation (4.123) was obtained by using the Leibnitz rule for the kth derivative of a product. Therefore,

$$y_k \equiv D^k y(x)|_{x=0} = c_1 + c_2 k, \tag{4.124}$$

which is easily shown to be the general solution of equation (4.122).

4.5. INHOMOGENEOUS EQUATIONS: METHOD OF UNDETERMINED COEFFICIENTS

We now turn to a technique for obtaining solutions to the nth-order linear inhomogeneous difference equation with constant coefficients,

$$y_{k+n} + a_1 y_{k+n-1} + \cdots + a_n y_k = R_k, \qquad a_n \neq 0, \tag{4.125}$$

when R_k is a linear combination of terms each having one of the forms

$$a^k, \qquad e^{bk}, \qquad \sin(ck), \qquad \cos(ck), \qquad k^l, \tag{4.126}$$

where a, b, and c are constants and l is a non-negative integer. We also include products of these forms; for example,

$$a^k \sin(ck), \qquad k^l e^{bk}, \qquad a^k k^l \cos(ck), \qquad \text{etc.} \tag{4.127}$$

If we allow for the possibility of complex values for the constant a, then each of the terms given in equations (4.126) and (4.127) is a particular instance of the term $k^l a^k$.

To proceed, we first need to introduce several definitions.

Definition: A *family* of a term R_k is the set of all functions of which R_k and $E^m R_k$, for $m = 1, 2, 3, \ldots$, are linear combinations.

Definition: A *finite family* is a family that contains only a finite number of functions.

For example, if $R_k = a^k$, then

$$E^m a^k = a^m a^k, \qquad m = 1, 2, 3, \ldots, \tag{4.128}$$

and the family of a^k contains only one member, namely, a^k. We denote this family by $[a^k]$.

If $R_k = k^l$, then

$$E^m k^l = (k+m)^l, \tag{4.129}$$

which can be expressed as a linear combination of $1, k, \ldots, k^l$; thus, the family of $E^m k^l$ is the set $[1, k, \ldots, k^l]$.

If $R_k = \cos(ck)$ or $\sin(ck)$, then the families are $[\cos(ck), \sin(ck)]$.

Finally, note that for the case where R_k is a product, the family consists of all possible products of the distinct members of the individual term families. For example, the term $R_k = k^l a^k$ has the finite family $[a^k, ka^k, \ldots, k^l a^k]$. Likewise, the term $R_k = k^l \cos(ck)$ has the finite family $[\cos(ck), k\cos(ck), \ldots, k^l \cos(ck), \sin(ck), k \sin(ck), \ldots, k^l \sin(ck)]$.

Definition: Let $g(E)$ be a polynomial operator function of the shift operator E. Let R_k be a given function of k, which can be expressed as a linear combination of terms having the form $k^l a^k$. Then $g(E)$ is said to be a nullifying operator if

$$g(E) R_k = 0; \tag{4.130}$$

that is, if the function R_k is a solution to the linear difference equation (4.130).

With these results in hand, we can now proceed to obtain particular solutions to equation (4.125) using the method of undetermined coefficients, where R_k consists of linear combinations of the family of the expression $k^l a^k$, where a can be a complex number.

First, assume that $g(E)$ is a nullifying operator of R_k on the right-hand side of equation (4.125). Applying $g(E)$ to both sides of equation (4.125) gives

$$g(E)f(E)y_k = 0, \tag{4.131a}$$

where

$$f(E) = E^m + a_1 E^{m-1} + \cdots + a_n. \tag{4.131b}$$

Therefore, we conclude that all solutions of the inhomogeneous equation (4.125) are included in the general solution of equation (4.131a), a homogeneous equation of higher order.

Let $f(r) = 0$ have n roots $r_1, r_2, \ldots, r_n$ and construct the function

$$y_k^{(H)} = c_1 y_k^{(1)} + c_2 y_k^{(2)} + \cdots + c_n y_k^{(n)}, \tag{4.132}$$

where the c_i are n arbitrary constants and the $y_k^{(i)}$ are a set of n linearly independent functions. Note that $y_k^{(H)}$ is the solution to the homogeneous equation

$$f(E)y_k = 0. \tag{4.133}$$

If $g(s) = 0$ has the t roots $s_1, s_2, \ldots, s_t$, then the characteristic equation of equation (4.131a) has $n + t$ roots $r_1, r_2, \ldots, r_n, s_1, s_2, \ldots, s_t$. Therefore, the general solution of equation (4.131a) will contain in addition to the terms in $y_k^{(H)}$, t new terms

$$D_1 v_k^{(1)} + D_2 v_k^{(2)} + \cdots + D_t v_k^{t}, \tag{4.134}$$

where the D_i are t constants. Furthermore, we see that the particular solution of equation (4.125) that does not contain the functions $y_k^{(i)}$ in $y_k^{(H)}$ is given by equation (4.134) when the *undetermined constants,* $D_1, D_2, \ldots, D_t$, are properly specified. To determine these constants, substitute

$$y_k^{(P)} = D_1 v_k^{(1)} + D_2 v_k^{(2)} + \cdots + D_t v_k^{(t)} \tag{4.135}$$

into the left-hand side of equation (4.125) and equate the coefficients of the same function of k on both sides.

There are two cases to consider.

Case I

None of the roots in the set $\{r_i\}$ occurs in the set $\{s_j\}$.

For this case, equation (4.135) is the general solution of the tth-order equation

$$g(E)y_k = 0. \tag{4.136}$$

The functions $v_k^{(i)}$, $i = 1, 2, \ldots, t$, are those that appear in R_k plus all those of the (finite) families of the individual terms that compose R.

Case II

Some of the roots in the set $\{r_i\}$ occur in the set $\{s_j\}$.

In this situation, the set $\{r_i\} + \{s_j\}$ now contains roots of higher multiplicity than the two individual sets of roots. To proceed, determine the general solution of equation (4.131a), drop all the functions $y_k^{(i)}$ that appear in $y_k^{(H)}$, and use the remaining functions to find the proper form for the particular solutions.

This procedure for obtaining particular solutions to the inhomogeneous equation (4.125) can be summarized as follows:

(i) Construct the family of R_k.

(ii) If the family contains no terms of the homogeneous solution, then write the particular solution $y_k^{(P)}$ as a linear combination of the members of that family. Determine the constants of combination such that the inhomogeneous difference equation is identically satisfied.

(iii) If the family contains terms of the homogeneous solution, then multiply each member of the family by the smallest integral power of k for which all such terms are removed. The particular solution $y_k^{(P)}$ can then be written as a linear combination of the members of this modified family. Again, determine the constants of combination such that the inhomogeneous difference equation is identically satisfied.

The following examples will illustrate the use of the method of undetermined coefficients.

4.5.1. Example A

The second-order difference equation

$$y_{k+2} - 5y_{k+1} + 6y_k = 2 + 4k \tag{4.137}$$

has the characteristic equation

$$r^2 - 5r + 6 = (r-3)(r-2) = 0, \qquad (4.138)$$

with roots $r_1 = 3$ and $r_2 = 2$. Therefore, the homogeneous solution is

$$y_k^{(H)} = c_1 3^k + c_2 2^k, \qquad (4.139)$$

where c_1 and c_2 are arbitrary constants. The right-hand side of equation (4.137) is $R_k = 4 + 4k$. Note that 2 has the family that consists of only one member [1], while $4k$ has the two-member family $[1, k]$. Therefore, the combined family is $[1, k]$. Since neither member of the combined family occurs in the homogeneous solution, we write the particular solution as the following linear combination:

$$y_k^{(P)} = A + Bk, \qquad (4.140)$$

where the constants A and B are to be determined.

Substitution of equation (4.140) into equation (4.137) gives

$$A + B(k+2) - 5A - 5B(k+1) + 6A + 6Bk = 2 + 4k. \qquad (4.141)$$

Upon setting the coefficients of the k^0 and k terms equal to zero, we obtain

$$2A - 3B = 2, \qquad 2B = 4. \qquad (4.142)$$

Therefore,

$$A = 4, \qquad B = 2, \qquad (4.143)$$

and the particular solution is

$$y_k^{(P)} = 4 + 2k. \qquad (4.144)$$

The general solution to equation (4.137) is

$$y_k = c_1 3^k + c_2 2^k + 4 + 2k. \qquad (4.145)$$

4.5.2. Example B

Consider the equation

$$y_{k+2} - 6y_{k+1} + 8y_k = 2 + 3k^2 - 5 \cdot 3^k. \qquad (4.146)$$

The characteristic equation is

$$r^2 - 6r + 8 = (r-2)(r-4) = 0, \tag{4.147}$$

which leads to the following solution of the homogeneous equation:

$$y_k^{(H)} = c_1 2^k + c_2 4^k, \tag{4.148}$$

where c_1 and c_2 are arbitrary constants. The families of the terms in R_k are

$$\begin{aligned} 2 &\rightarrow [1], \\ k^2 &\rightarrow [1, k, k^2], \\ 3^k &\rightarrow [3^k]. \end{aligned} \tag{4.149}$$

The combined family is $[1, k, k^2, 3^k]$ and contains no members that occur in the homogeneous solution. Therefore, the particular solution takes the form

$$y_k^{(P)} = A + Bk + Ck^2 + D3^k, \tag{4.150}$$

where A, B, C, and D are constants to be determined. Substitution of equation (4.150) into (4.146) and simplifying the resulting expression gives

$$(3A - 4B - 2C) + (3B - 8C)k + 3Ck^2 - D3^k = 2 + 3k^2 - 5 \cdot 3^k. \tag{4.151}$$

Equating the coefficients of the linearly independent terms on both sides to zero gives

$$3A - 4B - 2C = 2, \qquad 3B - 8C = 0, \qquad 3C = 3, \qquad D = 5, \tag{4.152}$$

which have the solution

$$A = 44/9, \qquad B = 8/3, \qquad C = 1, \qquad D = 5. \tag{4.153}$$

Consequently, the particular solution is

$$y_k^{(P)} = 44/9 + 8/3 k + k^2 + 5 \cdot 3^k, \tag{4.154}$$

and the general solution to equation (4.146) is

$$y_k = c_1 2^k + c_2 4^k + 44/9 + 8/3 k + k^2 + 5 \cdot 3^k. \tag{4.155}$$

4.5.3. Example C

The equation

$$y_{k+2} - 4y_{k+1} + 3y_k = k4^k \tag{4.156}$$

has the homogeneous solution

$$y_k^{(H)} = c_1 + c_2 3^k, \tag{4.157}$$

where c_1 and c_2 are arbitrary constants. The family of $R_k = k4^k$ is $[4^k, k4^k]$ and does not contain a term which appears in the homogeneous solution. Therefore, the particular solution is of the form

$$y_k^{(P)} = (A + Bk)4^k, \tag{4.158}$$

where the constants A and B can be determined by substituting equation (4.158) into equation (4.156); doing this gives

$$(3A + 16B)4^k + (3B)k4^k = k4^k \tag{4.159}$$

and

$$3A + 16B = 0, \qquad 3B = 1, \tag{4.160}$$

or

$$A = -16/9, \qquad B = 1/3. \tag{4.161}$$

The particular solution is

$$y_k^{(P)} = -16/9\, 4^k + 1/3\, k4^k, \tag{4.162}$$

and the general solution to equation (4.156) is

$$y_k = c_1 + c_2 3^k - 1/9(16 - 3k)4^k. \tag{4.163}$$

4.5.4. Example D

The second-order difference equation

$$y_{k+2} - 3y_{k+1} + 2y_k = 2 \sin (3k) \tag{4.164}$$

has the homogeneous solution

$$y_k^{(H)} = c_1 + c_2 2^k, \tag{4.165}$$

where c_1 and c_2 are arbitrary constants. The family of $R_k = 2\sin(3k)$ is $[\sin(3k), \cos(3k)]$ and contains no term which appears in the homogeneous solution. Therefore, the particular solution to equation (4.164) can be written

$$y_k^{(P)} = A\sin(3k) + B\cos(3k). \tag{4.166}$$

Substitution of this last equation into equation (4.164) and using the trigonometric relations

$$\begin{aligned} \sin(\theta_1 + \theta_2) &= \sin\theta_1 \cos\theta_2 + \cos\theta_1 \sin\theta_2, \\ \cos(\theta_1 + \theta_2) &= \cos\theta_1 \cos\theta_2 - \sin\theta_1 \sin\theta_2 \end{aligned} \tag{4.167}$$

gives

$$\begin{aligned} &\{A[\cos(6) - 3\cos(3) + 2] + B[-\sin(6) + 3\sin(3)]\}\sin(3k) \\ &\quad + \{A[\sin(6) - 3\sin(3)] + B[\cos(6) - 3\cos(3) + 2]\}\cos(3k) \\ &\qquad = 2\sin(3k). \end{aligned} \tag{4.168}$$

Therefore, A and B must satisfy the two linear equations

$$\begin{aligned} A[\cos(6) - 3\cos(3) + 2] + B[-\sin(6) + 3\sin(3)] &= 2, \\ A[\sin(6) + 3\sin(3)] + B[\cos(6) - 3\cos(3) + 2] &= 0. \end{aligned} \tag{4.169}$$

Evaluating the trigonometric functions (using the fact that their arguments are in radians) gives

$$\begin{aligned} \cos(3) &= -0.991, \qquad \sin(3) = 0.139, \\ \cos(6) &= 0.964, \qquad \sin(6) = -0.276. \end{aligned} \tag{4.170}$$

Thus, the solution to equation (4.169) for A and B is

$$A = 0.337, \qquad B = 0.039, \tag{4.171}$$

and the particular solution can be written

$$y_k^{(P)} = 0.337\sin(3k) + 0.039\cos(3k). \tag{4.172}$$

Finally, the general solution of equation (4.164) is

$$y_k = c_1 + c_2 2^k + 0.337 \sin(3k) + 0.039 \cos(3k). \tag{4.173}$$

4.5.5. Example E

Consider the third-order difference equation

$$y_{k+2} - 7y_{k+2} + 16y_{k+1} - 12y_k = k2^k. \tag{4.174}$$

Its characteristic equation is

$$r^3 - 7r^2 + 16r - 12 = (r-2)^2(r-3) = 0, \tag{4.175}$$

and the corresponding homogeneous solution is

$$y_k^{(H)} = (c_1 + c_2 k)2^k + c_3 3^k, \tag{4.176}$$

where c_1, c_2, and c_3 are arbitrary constants. The family of $R_k = k2^k$ is $[2^k, k2^k]$. Note that both members occur in the homogeneous solution; therefore, we must multiply the family by k^2 to obtain a new family which does not contain any function that appears in the homogeneous solution. The new family is $[k^2 2^k, k^3 2^k]$. Thus, the particular solution is

$$y_k^{(P)} = (Ak^2 + Bk^3)2^k, \tag{4.177}$$

where A and B are to be determined. The substitution of equation (4.177) into equation (4.174) gives

$$2^k(-8A + 24B) + k2^k(-24B) = k2^k, \tag{4.178}$$

or

$$8A - 24B = 0, \qquad -24B = 1, \tag{4.179}$$

and

$$A = -\tfrac{1}{8}, \qquad B = -\tfrac{1}{24}. \tag{4.180}$$

Therefore, the particular solution is

$$y_k^{(P)} = -\tfrac{1}{24}(3 + k)k^2 2^k, \tag{4.181}$$

and the general solution is

$$y_k = (c_1 + c_2 k)2^k + c_3 3^k - \tfrac{1}{24}(3 + k)k^2 2^k. \tag{4.182}$$

4.6. INHOMOGENEOUS EQUATIONS: OPERATOR METHODS

In this section, we present operator methods as another technique for obtaining particular solutions to linear inhomogeneous difference equations with constant coefficients.

The central idea of the technique is to start with the general nth-order linear inhomogeneous difference equation with constant coefficients, in the form

$$f(E)y_k = R_k, \tag{4.183}$$

and determine a particular solution by means of the relation

$$y_k^{(P)} = f(E)^{-1} R_k, \tag{4.184}$$

where $f(E)^{-1}$ is defined to be that operator such that when the right-hand side of equation (4.184) is operated upon by $f(E)$, we obtain R_k.

We proceed by stating and proving the following three theorems.

Theorem 4.5: Let $f(E)$ be a polynomial in E; then

$$f(E)a^k = f(a)a^k. \tag{4.185}$$

Proof: We have

$$f(E) = E^n + a_1 E^{n-1} + \cdots + a_n. \tag{4.186}$$

Therefore,

$$\begin{aligned} f(E)a^k &= (E^n + a_1 E^{n-1} + \cdots + a_n)a^k \\ &= a^{k+n} + a_1 a^{k+n-1} + \cdots + a_n a^k \\ &= a^k(a^n + a_1 a^{n-1} + \cdots + a_n) \\ &= a^k f(a). \end{aligned} \tag{4.187}$$

Theorem 4.6: Let $f(E)$ be a polynomial in E. Let F_k be a function of k. Then

$$f(E)a^k F_k = a^k f(aE)F_k. \tag{4.188}$$

Proof: Proceeding as in the previous theorem, we obtain

$$\begin{aligned} f(E)a^kF_k &= (E^n + a_1E^{n-1} + \cdots + a_n)a^kF_k \\ &= a^{k+n}F_{k+n} + a_1a^{k+n-1}F_{k+n-1} + \cdots + a_na^kF_k \\ &= a^k[a^nF_{k+n} + a_1a^{n-1}F_{k+n-1} + \cdots + a_nF_k] \\ &= a^k[(aE)^nF_k + a_1(aE)^{n-1}F_k + \cdots a_nF_k] \\ &= a^k[(aE)^n + a_1(aE)^{n-1} + \cdots + a_n]F_k = a^kf(aE)F_k. \end{aligned} \tag{4.189}$$

Theorem 4.7: Let F_k be a function of k. Then

$$(E-a)^ma^kF_k = a^ka^m(E-1)^mF_k = a^ka^m\Delta^mF^k. \tag{4.190}$$

Proof: This theorem is a special case of the preceding theorem.

Theorem 4.8: For $m = 1, 2, 3, \ldots,$ we have

$$(E-a)^{-m}a^k = \frac{k^{(m)}a^{k-m}}{m!}. \tag{4.191}$$

Proof: We will use mathematical induction to prove this theorem. For $m = 1$, we have

$$(E-a)^{-1}a^k = ka^{k-1} = \frac{k^{(1)}a^{k-1}}{1!}. \tag{4.192}$$

This result follows from

$$(E-a)ka^{k-1} = a^k. \tag{4.193}$$

Therefore, the result is true for $m = 1$. We now show that if the theorem is true for $m = p$, it is true for $m = p + 1$. Assume

$$(E-a)^{-p}a^k = \frac{k^{(p)}a^{k-p}}{p!}. \tag{4.194}$$

Hence,

$$\begin{aligned} (E-a)^{-p-1}a^k &= (E-a)^{-1}(E-a)^{-p}a^k \\ &= (E-a)^{-1}\frac{k^{(p)}a^{k-p}}{p!}. \end{aligned} \tag{4.195}$$

Now define the new function V_k as follows:

$$a^k V_k = (E - a)^{-1} \frac{k^{(p)} a^{k-p}}{p!}. \tag{4.196}$$

Now apply to both sides of equation (4.195) the operator $(E - a)$; doing this gives

$$(E - a) a^k V_k = \frac{k^{(p)} a^{k-p}}{p!}. \tag{4.197}$$

Therefore,

$$a^{k+1}(V_{k+1} - V_k) = \frac{k^{(p)} a^{k-p}}{p!}, \tag{4.198}$$

or

$$\Delta V_k = \frac{k^{(p)} a^{-p-1}}{p!}, \tag{4.199}$$

and

$$V_k = \frac{k^{(p+1)} a^{-p-1}}{p!}. \tag{4.200}$$

Substituting this result into equation (4.196) and using equation (4.195) gives

$$(E - a)^{-p-1} a^k = a^k V_k = \frac{k^{(p+1)} a^{k-p-1}}{(k+1)!} \tag{4.201}$$

Consequently, if the theorem is true for $m = p$, it is true for $m = p + 1$. Thus, by the method of induction, the theorem is true for all m.

Theorem 4.9: Let $f(a) \neq 0$. Then

$$f(E)^{-1} a^k = \frac{a^k}{f(a)}. \tag{4.202}$$

Proof: From Theorem 4.5, we have

$$f(E) a^k = f(a) a^k. \tag{4.203}$$

Applying $f(E)^{-1}$ to both sides of equation (4.203) gives

$$f(E)^{-1}f(E)a^k = f(a)f(E)^{-1}a^k, \tag{4.204}$$

or

$$a^k = f(a)f(E)^{-1}a^k. \tag{4.205}$$

Therefore, on division by $f(a)$, we obtain equation (4.202).

Theorem 4.9′: Let $f(a) = 0$, where $f(E) = (E - a)^m g(E)$ and $g(a) \neq 0$. Then

$$f(E)^{-1}a^k = \frac{a^{k-m}k^m}{g(a)m!}. \tag{4.206}$$

Proof: We assume that $f(r)$ has a zero at $r = a$ of multiplicity m. Therefore,

$$\begin{aligned} f(E)^{-1}a^k &= (E - a)^{-m}g(E)^{-1}a^k = (E - a)^{-m}\frac{a^k}{g(a)} \\ &= \frac{a^k}{g(a)}(aE - a)^{-m} \cdot 1 = \frac{a^{k-m}}{g(a)}(E - 1)^{-m} \cdot 1 \qquad (4.207) \\ &= \frac{a^{k-m}}{g(a)}\Delta^{-m} \cdot 1 = \frac{a^{k-m}}{g(a)}\frac{k^{(m)}}{m!}. \end{aligned}$$

This is just the result of equation (4.206).

The following two results are of interest.

Result 1: Let P_k be a polynomial function in k of degree m and define u_k to be

$$u_k = f(E)^{-1}P_k = \frac{1}{f(E)}P_k. \tag{4.208}$$

Now

$$\Delta^{m+1}P_k = 0. \tag{4.209}$$

Therefore, if we expand the operator

$$\frac{1}{f(E)}=\frac{1}{f(1+\Delta)}, \tag{4.210}$$

$$\frac{1}{f(1+\Delta)}=\frac{1}{f(1)}+b_1\Delta+b_2\Delta^2+\cdots+b_i\,\Delta^i+\cdots \tag{4.211}$$

in an ascending series of powers of Δ, then we can stop at the term in Δ^m. Hence

$$u_k=\frac{1}{f(1+\Delta)}P_k=\left(\frac{1}{f(1)}+b_1\Delta+b_2\Delta^2+\cdots+b_m\,\Delta^m\right)P_k. \tag{4.212}$$

Result 2: Consider a^kP_k, where P_k is a polynomial function in k of degree m. Define w_k to be

$$w_k=f(E)^{-1}a^kP_k. \tag{4.213}$$

Now let us show that

$$f(E)^{-1}a^kP_k=a^kf(aE)^{-1}P_k. \tag{4.214}$$

Operate on both left- and right-hand sides of this equation with $f(E)$ to obtain

$$f(E)f(E)^{-1}a^kP_k=a^kP_k, \tag{4.215}$$

and

$$f(E)[a^kf(aE)^{-1}P_k-=a^kf(aE)[f(aE)^{-1}P_k]=a^kP_k. \tag{4.216}$$

If we replace E by $1+\Delta$, then equation (4.213) becomes

$$\begin{aligned} w_k&=\frac{1}{f(1+\Delta)}a^kP_k=a^k\frac{1}{f(a+a\Delta)}P_k\\ &=a^k\left(\frac{1}{f(a)}+b_1\Delta+b_2\Delta^2+\cdots+b_m\,\Delta^m\right)P_k. \end{aligned} \tag{4.217}$$

We can use Theorems 4.5 to 4.9′ and Results 1 and 2 to obtain particular solutions to linear, inhomogeneous difference equations having constant coefficients. In general, for practical calculations, this technique is limited to inhomogeneous terms that correspond to linear combinations of members of the family of the function $k^m a^k$, where a can be a complex number.

The following problems illustrate the use of the method of operators.

4.6.1. Example A

The second-order inhomogeneous equation

$$y_{k+2} - y_{k+1} - 2y_k = 6 \tag{4.218}$$

can be written in the operator form

$$(E + 1)(E - 2)y_k = 6.$$

Thus, we see immediately that the homogeneous solution is

$$y_k^{(H)} = c_1(-1)^k + c_2 2^k, \tag{4.219}$$

where c_1 and c_2 are arbitrary constants. The particular solution is given by the expression

$$y_k^{(P)} = [(E + 1)(E - 2)]^{-1}6. \tag{4.220}$$

Using equation (4.212), we obtain

$$y_k^{(P)} = \frac{6}{(2)(-1)} = -3. \tag{4.221}$$

Therefore, the general solution of equation (4.218) is

$$y_k = c_1(-1)^k + c_2 2^k - 3. \tag{4.222}$$

4.6.2. Example B

The equation

$$y_{k+2} - 3y_{k+1} + 2y_k = (E - 1)(E - 2)y_k = 1 \tag{4.223}$$

has the homogeneous solution

$$y_k^{(H)} = c_1 + c_2 2^k, \tag{4.224}$$

where c_1 and c_2 are arbitrary constants.

Now $f(E) = (E - 1)(E - 2)$ and $f(1) = 0$, where the zero is simple. Using Theorem 4.9′, we find

$$y_k^{(P)} = [(E-1)(E-2)]^{-1} \cdot 1 = \frac{k}{(1-2)(1!)} = -k. \tag{4.225}$$

Thus, the general solution is

$$y_k = c_1 + c_2 2^k - k. \tag{4.226}$$

4.6.3. Example C

The equation

$$y_{k+2} - 5y_{k+1} + 6y_k = 3^k \tag{4.227}$$

can be expressed as

$$f(E)y_k = (E-2)(E-3)y_k = 3^k. \tag{4.228}$$

The particular solution is

$$y_k^{(P)} = (E-3)^{-1}(E-2)^{-1}3^k. \tag{4.229}$$

Note that $f(3) = 0$; applying Theorem 4.9′, we obtain

$$y_k^{(P)} = \frac{k3^{k-1}}{(3-2)(1!)} = k3^{k-1}. \tag{4.230}$$

Therefore, the general solution of equation (4.227) is

$$y_k = c_1 2^k + c_2 3^k + k3^{k-1}. \tag{4.231}$$

4.6.4. Example D

Consider the following equation:

$$f(E)y_k = (E-2)^2 y_k = 2k.$$

The solution to the homogeneous equation is

$$y_k^{(H)} = (c_1 + c_2 k)2^k. \tag{4.232}$$

Now $f(E)$ has a double root and we must apply Theorem 4.9′ again. [For this case $g(E) = 1$ and $m = 2$.] Thus,

$$y_k^{(P)} = (E-2)^{-2} \cdot 2k = \frac{2^{k-2}k^2}{(1)(2!)} = \frac{k^2 2^k}{8}, \tag{4.233}$$

and the general solution of equation (4.232) is

$$y_k = (c_1 + c_2 k)2^k + \frac{k^2 2^k}{8}. \tag{4.234}$$

4.6.5. Example E

The third-order equation

$$y_{k+3} - y_{k+2} - 4y_{k+1} + 4y_k = 1 + k + 2^k \tag{4.235}$$

can be expressed in the operator form

$$f(E)y_k = (E-1)(E-2)(E+2)y_k = 1 + k + 2^k. \tag{4.236}$$

Consequently, the homogeneous solution is

$$y_k^{(P)} = c_1 + c_2 2^k + c_3(-2)^k, \tag{4.237}$$

where c_1, c_2, and c_3 are constants. The particular solution will be determined by separating the right-hand side of equation (4.236) into two terms, $1 + k$ and 2^k, and letting $f(E)^{-1}$ act on each. We have

$$\begin{aligned}
&[(E-1)(E-2)(E+2)]^{-1}(1+k) \\
&\qquad = -[\Delta(1-\Delta)(3+\Delta)]^{-1}(1+k) \\
&\qquad = \left(-\frac{1}{3}\right)\frac{\Delta^{-1}}{(1-\Delta)(1+\tfrac{1}{3}\Delta)}(1+k) \\
&\qquad = (-\tfrac{1}{3})\Delta^{-1}(1+\Delta+\cdots)(1-\tfrac{1}{3}\Delta+\cdots)(1+k) \qquad (4.238) \\
&\qquad = (-\tfrac{1}{3})\Delta^{-1}(1+\tfrac{2}{3}\Delta)(1+k) \\
&\qquad = (-\tfrac{1}{3})\Delta^{-1}(\tfrac{5}{3}+k) \\
&\qquad = (-\tfrac{1}{3})[\tfrac{5}{3}k + \tfrac{1}{2}k(k-1)] \\
&\qquad = (-\tfrac{7}{9})k - \tfrac{1}{6}k^2.
\end{aligned}$$

Lines two and three on the right-hand side of equation (4.238) follow from equations (4.209) and (4.212).

Also, from Theorem 4.9′, we have

$$[(E-2)(E-1)(E+2)]^{-1}2^k = \frac{k2^{k-1}}{(1\cdot 4)(1!)} = \frac{k2^k}{8}. \tag{4.239a}$$

Therefore, the particular solution is

$$y_k^{(P)} = -\tfrac{7}{9}k - \frac{k^2}{6} + \frac{k2^k}{8}; \tag{4.239b}$$

consequently, the general solution of equation (4.235) is given by the expression

$$y_k = c_1 + c_2 2^k + c_3(-2)^k - \frac{7}{9}k - \frac{k^2}{6} + \frac{k2^k}{8}. \tag{4.240}$$

4.6.6. Example F

Consider the equation

$$f(E)y_k = (E-2)(E-3)y_k = (5-k+k^2)4^k. \tag{4.241}$$

The homogeneous solution is

$$y_k^{(H)} = c_1 2^k + c_2 3^k, \tag{4.242}$$

where c_1 and c_2 are arbitrary constants. The particular solution is given by

$$y_k^{(P)} = [(E-2)(E-3)]^{-1}(5-k+k^2)4^k. \tag{4.243}$$

Now, using equation (4.214), we obtain

$$y_k^{(P)} = 4^k[(4E-2)(4E-3)]^{-1}(5-k+k^2), \tag{4.244}$$

and on setting $E = 1 + \Delta$,

$$\begin{aligned} y_k^{(P)} &= \tfrac{1}{2}4^k(1+6\Delta+8\Delta^2)^{-1}(5-k+k^2) \\ &= \tfrac{1}{2}4^k(1-6\Delta+28\Delta^2+\cdots)(5-k+k^2) \\ &= \tfrac{1}{2}4^k[5-k+k^2-6(2k)+28(2)] \\ &= \tfrac{1}{2}4^k(61-13k+k^2). \end{aligned} \tag{4.245}$$

Thus, the general solution of equation (4.241) is

$$y_k = c_1 2^k + c_3 3^k + \tfrac{1}{2} 4^k (61 - 13k + k^2). \tag{4.246}$$

4.7. *z*-TRANSFORM METHOD

The z-transform is a mathematical operation which transforms functions of a discrete variable k into functions of the continuous transform variable z.

In this section, we show how the z-transform method can be used to obtain solutions to linear difference equations with constant coefficients. The method has the property of transforming linear difference equations into algebraic equations which can then be handled by using the familiar laws of algebra.

Definition: Let $\{y_k\}$ be a sequence of numbers such that $y_k = 0$ for $k < 0$. The z-transform of this sequence is the series

$$z(y_k) = \sum_{k=0}^{\infty} \frac{y_k}{z^k}, \tag{4.247}$$

where z is the transform variable.

Note that $Z(y_k)$ is a function of z.

As far as the convergence of the series, given by equation (4.247), is concerned, there are two approaches that can be taken. First, we can take the above definition in a formal sense without regard for conditions of convergence. Consequently, the variable z is regarded simply as a symbol and the transform $Z(y_k)$ as a series that is never actually summed, but can be manipulated algebraically. Second, we can view the series representation of $Z(y_k)$ as defining an actual function of the variable z, where numerical values of z (which we can take to be complex) give numerical values for $Z(y_k)$. Thus, $Z(y_k)$ is taken to be a complex function of the complex variable z. From this point of view, one must consider the conditions for which the series, defined by equation (4.247), converges. As it turns out, this is rather easy to discover since in all our applications we have, for some (real) constant $c \geq 0$, the condition

$$|y_k| \leq c^k. \tag{4.248}$$

Using this result and the d'Alembert ratio test for convergence, we conclude that $Z(y_k)$ converges for sufficiently large z. In detail, the condition is

$$c < |z|. \tag{4.249}$$

We now turn to a study of the main properties of the z-transform; they are summarized in the following eight theorems. While we will not use all the results given by the theorems to follow, the results are quite interesting. Those with a knowledge of differential equations and the Laplace transform will immediately see the close parallels between the properties of linear differential and difference equations having constant coefficients.

Theorem 4.10: Let A and B be constants. Let $\{y_k\}$ and $\{w_k\}$ be sequences with the property that $y_k = 0$ and $w_k = 0$ for $k < 0$. Then

$$Z(Ay_k + Bw_k) = AZ(y_k) + BZ(w_k). \tag{4.250}$$

Proof: We have

$$\begin{aligned} Z(Ay_k + Bw_k) &= \sum_{k=0}^{\infty} (Ay_k + Bw_k) z^{-k} \\ &= A \sum_{k=0}^{\infty} y_k z^{-k} + B \sum_{k=0}^{\infty} w_k z^{-k} \\ &= AZ(y_k) + BZ(w_k). \end{aligned} \tag{4.251}$$

The z-transform is thus a linear transformation.

Theorem 4.11: If $n > 0$, then

$$Z(y_{k+n}) = z^n Z(y_k) - \sum_{m=0}^{n-1} y_m z^{n-m}. \tag{4.252}$$

Proof: From the definition of the z-transform, we have

$$\begin{aligned} Z(y_{k+n}) &= \sum_{k=0}^{\infty} y_{k+n} z^{-k} \\ &= z^n \sum_{k=0}^{\infty} y_{k+n} z^{-k-n} \\ &= z^n \left(\sum_{k=0}^{\infty} y_k z^{-k} - \sum_{k=0}^{n-1} y_k z^{-k} \right) \\ &= z^n Z(y_k) - \sum_{m=0}^{n-1} y_m z^{n-m}, \end{aligned} \tag{4.253}$$

which is just equation (4.252).

Theorem 4.12: If $k > 0$, then

$$Z(y_{k-n}) = z^{-n}Z(y_k). \tag{4.254}$$

Proof: By definition, we have

$$Z(y_{k-n}) = \sum_{k=0}^{\infty} y_{k-n}z^{-k} = z^{-n}\sum_{k=0}^{\infty} y_{k-n}z^{-(k-n)}. \tag{4.255}$$

Now let $l = k - n$; therefore, for $k = 0$, $l = -n$ and equation (4.255) becomes

$$Z(y_{k-n}) = z^{-n}\sum_{l=-n}^{\infty} y_l z^{-l}. \tag{4.256}$$

However, $y_l = 0$ for $l < 0$; therefore,

$$Z(y_{k-n}) = z^{-n}\sum_{l=0}^{\infty} y_l z^{-l} = z^{-n}Z(y_k), \tag{4.257}$$

which is the result given by equation (4.254).

Theorem 4.13: The z-transform of the finite sum of a sequence is

$$Z\left(\sum_{i=0}^{k} y_i\right) = \frac{z}{z-1}Z(y_k). \tag{4.258}$$

Proof: Let

$$g_k = \sum_{i=0}^{k} y_i. \tag{4.259}$$

Therefore,

$$g_{k-1} = \sum_{i=0}^{k-1} y_i \tag{4.260}$$

and

$$y_k = g_k - g_{k-1}. \tag{4.261}$$

Using Theorem 4.12, we obtain from equation (4.261)

$$Z(y_k) = \frac{z-1}{z} Z\left(\sum_{i=0}^{k} y_i\right), \tag{4.262}$$

which, after rearranging factors, is just equation (4.258).

Theorem 4.14: The z-transform of ky_k is

$$Z(ky_k) = -z\frac{d}{dz}[Z(y_k)]. \tag{4.263}$$

Proof: By definition

$$\begin{aligned} Z(ky_k) &= \sum_{k=0}^{\infty} ky_k z^{-k} = -z\sum_{k=0}^{\infty} y_k(-kz^{-k-1}) \\ &= -z\sum_{k=0}^{\infty} y_k \frac{d}{dz}(z^{-k}) = -z\frac{d}{dz}\left(\sum_{k=0}^{\infty} y_k z^{-k}\right) \\ &= -z\frac{d}{dz}[Z(y_k)]. \end{aligned} \tag{4.264}$$

Theorem 4.15: Let $F(z)$ denote the z-transform of y_k,

$$F(z) = Z(y_k). \tag{4.265}$$

Then the z-transform of $a^k y_k$ is

$$Z(a^k y_k) = F\left(\frac{z}{a}\right). \tag{4.266}$$

Proof: By definition

$$Z(a^k y_k) = \sum_{k=0}^{\infty} a^k y_k z^{-k} = \sum_{k=0}^{\infty} y_k \left(\frac{a}{z}\right)^k = F\left(\frac{z}{a}\right). \tag{4.267}$$

Theorem 4.16: Let $F(z)$ denote the z-transform of y_k; then if the limit of $F(z)$ as $z \to \infty$ exists, the following relation is true:

$$\lim_{k\to 0} y_k = \lim_{z\to\infty} F(z). \tag{4.268}$$

Proof: From the definition of the *z*-transform

$$F(z) = Z(y_k) = \sum_{k=0}^{\infty} y_k z^{-k} = y_0 + \frac{y_1}{z} + \frac{y_2}{z^2} + \cdots. \tag{4.269}$$

Taking the limit as $z \to \infty$, we obtain the result given by (4.268).

Theorem 4.17: If $(z - 1)F(z)$, where $F(z)$ is the *z*-transform of y_k, is analytic for $z \geq 1$, then

$$\lim_{k\to\infty} y_k = \lim_{z\to 1} (z - 1)F(z). \tag{4.270}$$

Proof: The *z*-transform can be written

$$Z(y_k) = \lim_{n\to\infty} \sum_{k=0}^{n} y_k z^{-k}. \tag{4.271}$$

Therefore

$$Z(y_{k+1} - y_k) = \lim_{n\to\infty} \sum_{k=0}^{n} (y_{k+1} - y_k) z^{-k}. \tag{4.272}$$

Using Theorem 4.11, we obtain the following result:

$$(z - 1)F(z) - zy_0 = \lim_{n\to\infty} \sum_{k=0}^{n} (y_{k+1} - y_k) z^{-k}. \tag{4.273}$$

Letting $z \to 1$ on both sides of equation (4.273) gives

$$\begin{aligned} \lim_{z\to 1} (z - 1)F(z) - y_0 &= \lim_{n\to\infty} \sum_{k=0}^{n} (y_{k+1} - y_k) \\ &= \lim_{n\to\infty} y_{n+1} - y_0 \end{aligned} \tag{4.274}$$

or

$$\lim_{z\to 1} (z - 1)F(z) = \lim_{k\to\infty} y_k, \tag{4.275}$$

which is the result given by equation (4.270).

Table 4.1
Selected z-Transform Pairs

NUMBER	y_k	$F(z) = Z(y_k)$
1	1	$\frac{1}{z-1}$
2	k	$\frac{z}{(z-1)^2}$
3	k^2	$\frac{z(z+1)}{(z-1)^3}$
4	a^k	$\frac{z}{z-a}$
5	ka^k	$\frac{az}{(z-a)^2}$
6	k^2a^k	$\frac{az(z+a)}{(z-a)^3}$
7	$\sin(ak)$	$\frac{z \sin a}{z^2 - 2z \cos a + 1}$
8	$\cos(ak)$	$\frac{z^2 - z \cos a}{z^2 - 2z \cos a + 1}$
9	$b^k \sin(ak)$	$\frac{bz \sin a}{z^2 - 2bz \cos a + b^2}$
10	$b^k \cos(ak)$	$\frac{z^2 - bz \cos a}{z^2 - 2bz \cos a + b^2}$

Given a sequence $\{y_k\}$ defined for $k \geq 0$, the corresponding z-transform can be determined by using the definition given in equation (4.247). Table 4.1 gives a short list of selected z-transform pairs. The examples presented at the end of this section illustrate how the various z-transforms can be calculated.

An inspection of Table 4.1 shows that, in general, the z-transform of sequences whose general member is a linear combination of terms having the form $k^m a^k$, where $m \geq 0$ and a can be complex, is a rational function of z. The following definitions are needed before we can proceed.

Definition: Let $F(z)$ be a function of the variable z; $F(z)$ is a rational function of z if it can be written as the ratio of two polynomial functions of z,

$$F(z) = \frac{b_0 z^m + b_1 z^{n-1} + \cdots + b_m}{z^n + a_1 z^{n-1} + \cdots + a_n}, \tag{4.276}$$

where $b_0, \ldots, b_m$ and $a_1, \ldots, a_n$ are given constants with $b_0 \neq 0$.

Definition: A rational function of z is called proper if the degree of the numerator polynomial is equal to or less than the degree of the denominator polynomial, i.e., $m \leq n$. It is strictly proper if $m < n$.

Definition: If for a rational function of z the numerator and denominator polynomials have no common factors, then the function is said to be reduced.

Definition: The degree of a rational function is the degree of its denominator polynomial when the function is in reduced form.

The following theorem shows that there is a direct relationship between sequences generated by linear, homogeneous difference equations with constant coefficients and proper rational z-transform functions.

Theorem 4.18: A sequence $\{y_k\}$, $k = 0, 1, 2, \ldots$, has a z-transform $Z(y_k) = F(z)$ that can be represented as a reduced proper rational function of degree n if and only if there exist n constants $a_1, a_2, \ldots, a_n$ such that

$$y_{k+n} + a_1 y_{k+n-1} + \cdots + a_n y_k = 0, \tag{4.277}$$

for all $k = 0, 1, 2, \ldots$.

Proof: Assume that y_k has a reduced proper rational z-transform of degree n. This means that $F(z)$ can be written as

$$\begin{aligned} F(z) &= \frac{b_0 z^n + b_1 z^{n-1} + \cdots + b_n}{z^n + a_1 z^{n-1} + \cdots + a_n} \\ &= y_0 + y_1 z^{-1} + y_2 z^{-2} + \cdots. \end{aligned} \tag{4.278}$$

Multiplying this equation by the denominator polynomial gives

$$\begin{aligned} &b_0 z^n + b_1 z^{n-1} + \cdots + b_n \\ &\quad = (z^n + a_1 z^{n-1} + \cdots + a_n)(y_0 + y_1 z^{-1} + y_2 z^{-2} + \cdots). \end{aligned} \tag{4.279}$$

If the coefficients of the same powers of z on the left- and right-hand sides of this equation are set equal, then the following relations are obtained:

$$
\begin{aligned}
b_0 &= y_0, \\
b_1 &= a_1 y_0 + y_1, \\
b_2 &= a_2 y_0 + a_1 y_1 + y_2, \\
&\vdots \\
b_n &= a_n y_0 + a_{n-1} y_1 + \cdots + a_1 y_{n-1} + y_n,
\end{aligned}
\tag{4.280a}
$$

and

$$
0 = a_n y_k + a_{n-1} y_{k+1} + \cdots + a_1 y_{k+n-1} + y_{k+n}. \tag{4.280b}
$$

The last relation follows from the fact that on the left-hand side of equation (4.279) the coefficient of z^{-k} is zero for $k > 0$. Thus, we conclude that for the z-transform function given by equation (4.278), the sequence $\{y_k\}$ must satisfy the nth-order linear homogeneous difference equation (4.277).

If now we start with equation (4.280b) and choose constants $b_0, b_1, \ldots, b_n$ such that equations (4.280a) are satisfied, then it is easy to show that the corresponding $F(z)$ is equal to equation (4.278).

The z-transform provides another method for obtaining solutions of linear difference equations with constant coefficients. The procedure is to take the z-transform of each term in the difference equation and solve for the z-transform $F(z)$ of y_k. In general, an algebraic equation is obtained for $F(z)$. Once this has been done, the inverse transformation back to y_k gives the desired solution of the difference equation.

The inverse transformation can be done in a number of ways, including contour integration, tables of transform pairs, power-series expansions, Maclaurin-series expansion, and partial-fraction expansion. For most problems, where the z-transform is a rational function, the method of partial-fraction expansion is both the easiest and the most straightforward to carry out.

We now show how the z-transform method can be used to determine solutions to the general nth-order inhomogeneous difference equation with constant coefficients,

$$
y_{k+n} + a_1 y_{k+n-1} + \cdots + a_{n-1} y_{k+1} + a_n y_k = R_k, \tag{4.281}
$$

where R_k is a given function of k. This expression can be written

$$
\sum_{i=0}^{n} a_i y_{k+n-i} = R_k, \qquad a_0 = 1. \tag{4.282}
$$

From Theroem 4.11, equation (4.252), we have

$$Z(y_{k+n-i}) = z^{n-i}F(z) - \sum_{p=0}^{n-i-1} y_p z^{n-i-p}, \tag{4.283}$$

where $F(z) = Z(y_k)$. Now define the z-transform of R_k to be

$$G(z) = Z(R_k). \tag{4.284}$$

Therefore, taking the z-transform of each term in equation (4.282) gives

$$\sum_{i=0}^{n} a_i \left(z^{n-i}F(z) - \sum_{p=0}^{n-i-1} y_p z^{n-i-p} \right) = G(z), \tag{4.285}$$

which can be solved for $F(z)$:

$$F(z) = \frac{G(z) + \sum_{i=0}^{n} a_i \sum_{p=0}^{n-i-1} y_p z^{n-i-p}}{z^n + a_1 z^{n-1} + \cdots + a_n}. \tag{4.286}$$

Note that the denominator is just the characteristic equation.

If R_k consists of linear combinations of the expressions a^k, k^m, $\cos(bk)$, $\sin(bk)$, and their products, then its z-transform will be a rational function. In this case, $F(z)$ can always be written as a partial-fraction expansion. The individual terms in this expansion can then be inverted to give its contribution to the solution y_k. The easiest procedure for carrying out this inverse transformation is to refer to a table of transform pairs.

It should be pointed out that the solution to an nth-order difference equation obtained by the z-transform method expresses the solution in terms of the given n initial conditions $y_0, y_1, \ldots, y_n$.

We now illustrate the use of the z-transform in the following examples.

4.7.1 Example A

Let us calculate the z-transform of $y_k = 1$. We have

$$\begin{aligned} F(z) = Z(1) &= 1 + z^{-1} + z^{-2} + z^{-3} + \cdots \\ &= \frac{1}{1 - z^{-1}}, \quad \text{for } |z| > 1, \\ &= \frac{z}{z-1}. \end{aligned} \tag{4.287}$$

4.7.2. Example B

The sequence $y_k = a^k$, where a is a constant, has the z-transform

$$\begin{aligned} F(z) &= Z(a^k) = 1 + az^{-1} + a^2 z^{-2} + a^3 z^{-3} + \cdots \\ &= \frac{1}{1 - az^{-1}}, \qquad \text{for } |z| > a, \\ &= \frac{z}{z - a}. \end{aligned} \tag{4.288}$$

4.7.3. Example C

Consider the sequence defined as follows: $y_0 = 4$, $y_1 = -1$, $y_2 = 2$, $y_k = 0$, for $k \geq 3$. The corresponding z-transform is

$$F(z) = 4 - \frac{1}{z} + \frac{2}{z^2} = \frac{4z^2 - z + 2}{z^2}. \tag{4.289}$$

4.7.4. Example D

The sequence $y_k = k$ has the z-transform

$$\begin{aligned} F(z) &= z^{-1} + 2z^{-2} + 3z^{-3} + 4z^{-4} + \cdots \\ &= z^{-1}(1 + 2z^{-1} + 3z^{-2} + 4z^{-3} + \cdots). \end{aligned} \tag{4.290}$$

Using the fact that

$$\frac{1}{(1-x)^2} = 1 + 2x + 3x^2 + 4x^2 + \cdots, \qquad |x| < 1, \tag{4.291}$$

we obtain

$$F(z) = z^{-1}(1 - z^{-1})^{-2}, \tag{4.292}$$

or

$$F(z) = Z(k) = \frac{z}{(z-1)^2}. \tag{4.293}$$

4.7.5. Example E

Consider the first-order equation

$$y_{k+1} - ay_k = R_k, \tag{4.294}$$

where a is a constant. Taking the z-transform of each term in this equation gives

$$zF(z) - zy_0 - aF(z) = G(z), \tag{4.295}$$

where $G(z) = z(R_k)$. Solving for $F(z)$, we obtain

$$F(z) = \frac{G(z) + zy_0}{z - a}. \tag{4.296}$$

Suppose $R_k = 0$; then $G(z) = 0$ and equatiion (4.296) becomes

$$F(z) = \frac{y_0 z}{z - a}. \tag{4.297}$$

From Table 4.1, we see that the inverse transform of $z/(z - a)$ is a^k. Therefore, the general solution to the difference equation

$$y_{k+1} - ay_k = 0 \tag{4.298}$$

is

$$y_k = y_0 a^k, \tag{4.299}$$

as can be readily checked by using some other method of solution. Note that there is one arbitrary constant, namely, y_0.

Now suppose $R_k = 1$, with $a \neq 1$. Therefore, from equation (4.296), we have

$$F(z) = \frac{z/(z - 1) + zy_0}{z - a} = \frac{y_0 z^2 + (1 - y_0)z}{(z - 1)(z - a)}, \tag{4.300}$$

where

$$G(z) = Z(1) = \frac{z}{z - 1}. \tag{4.301}$$

Therefore, writing

$$\frac{F(z)}{z} = \frac{A}{z-1} + \frac{B}{z-a} = \frac{z(A+B)-(aA+B)}{(z-1)(z-a)} \tag{4.302}$$

and comparing with equation (4.300) gives the following two equations for A and B:

$$\begin{aligned} A + B &= y_0, \\ aA + B &= y_0 - 1. \end{aligned} \tag{4.303}$$

Solving gives

$$A = \frac{1}{1-a}, \qquad B = \frac{(1-a)y_0 - 1}{1-a} \tag{4.304}$$

and

$$F(z) = \frac{1}{1-a}\frac{z}{z-1} + \frac{(1-a)y_0 - 1}{1-a}\frac{z}{z-a}. \tag{4.305}$$

From Table 4.1, we determine the inverse transform of equation (4.305) to be

$$y_k = \frac{1}{1-a} + \frac{(1-a)y_0 - 1}{1-a} a^k. \tag{4.306}$$

This is the general solution to the difference equation

$$y_{k+1} - ay_k = 1. \tag{4.307}$$

Finally, let $R_k = b^k$, $b \neq a$. We have

$$G(z) = Z(b^k) = \frac{z}{z-b}, \tag{4.308}$$

and from equation (4.296)

$$F(z) = \frac{z/(z-b) + zy_0}{z-a}, \tag{4.309}$$

or

$$\frac{F(z)}{z} = \frac{y_0 z + (1 - b y_0)}{(z-a)(z-b)}. \tag{4.310}$$

Partial-fraction expansion of the right-hand side of equation (4.310) gives

$$\frac{F(z)}{z} = \frac{1}{b-a}\frac{z}{z-b} + \frac{y_0(b-a)-1}{b-a}\frac{z}{z-a}. \tag{4.311}$$

Carrying out the inverse transformation gives

$$y_k = \frac{1}{b-a} b^k + \frac{y_0(b-a)-1}{b-a} a^k. \tag{4.312}$$

4.7.6. Example F

Taking the z-transform of each term of the second-order equation

$$y_{k+2} + a_1 y_{k+1} + a_2 y_k = R_k \tag{4.313}$$

gives

$$[z^2 F(z) - z^2 y_0 - z y_1] + a_1[zF(z) - z y_0] + a_2 F(z) = G(z), \tag{4.314}$$

or

$$F(z) = \frac{G(z) + y_0 z^2 + (a_1 y_0 + y_1)}{z^2 + a_1 z + a_2}, \tag{4.315}$$

where $G(z) = Z(R_k)$.

Let $a_1 = 0$, $a_2 = 4$, and $R_k = 0$. This corresponds to the difference equation

$$y_{k+2} - 4y_k = 0. \tag{4.316}$$

The z-transform is, from equation (4.315), given by the expression

$$\frac{F(z)}{z} = \frac{z y_0 + y_1}{(z+2)(z-2)} = \frac{2y_0 - y_1}{4}\frac{1}{z+2} + \frac{2y_0 + y_1}{4}\frac{1}{z-2}. \tag{4.317}$$

From Table 4.1, we can obtain the inverse transform of $F(z)$; doing this gives

$$y_k = \tfrac{1}{4}(2y_0 - y_1)(-2)^k + \tfrac{1}{4}(2y_0 + y_1)2^k. \tag{4.318}$$

Since y_0 and y_1 are arbitrary, this gives the general solution of equation (4.316).

The equation

$$y_{k+2} + 4y_{k+1} + 3y_k = 2^k \tag{4.319}$$

corresponds to $a_1 = 4$, $a_2 = 3$, and $R_k = 2^k$. Using the fact that

$$G(z) = Z(2^k) = \frac{z}{z-2}, \tag{4.320}$$

we can rewrite equation (4.315), for this case, as

$$\frac{F(z)}{z} = \frac{y_0(z+4)}{(z+1)(z+3)} + \frac{y_1}{(z+1)(z+3)} + \frac{1}{(z-2)(z+1)(z+3)}. \tag{4.321}$$

The three terms on the right-hand side of this equation have the following partial-fraction expansions:

$$\frac{1}{(z+1)(z+3)} = \frac{1}{2}\frac{1}{z+1} - \frac{1}{2}\frac{1}{z+3}, \tag{4.322}$$

$$\frac{z+4}{(z+1)(z+3)} = \frac{3}{2}\frac{1}{z+1} - \frac{1}{2}\frac{1}{z+3}, \tag{4.323}$$

$$\frac{1}{(z-2)(z+1)(z+3)} = \frac{1}{15}\frac{1}{z-2} - \frac{1}{6}\frac{1}{z+1} + \frac{1}{10}\frac{1}{z+3}. \tag{4.324}$$

Therefore,

$$\frac{F(z)}{z} = \frac{1}{6}\frac{3y_1 + 9y_0 - 1}{z+1} + \frac{1}{10}\frac{1 - 5y_0 - 5y_1}{z+3} + \frac{1}{15}\frac{1}{z-2}, \tag{4.325}$$

and taking the inverse transform gives

$$y_k = \tfrac{1}{6}(3y_1 + 9y_0 - 1)(-1)^k + \tfrac{1}{10}(1 - 5y_0 - 5y_1)(-3)^k + \tfrac{1}{15}2^k. \tag{4.326}$$

4.8. SYSTEMS OF DIFFERENCE EQUATIONS

In this section, we consider systems of simultaneous difference equations having constant coefficients. These types of systems arise quite naturally, for we can always write the nth-order equation

$$y_{k+n} + a_1 y_{k+n-1} + \cdots + a_{n-1} y_{k+1} + a_n y_k = R_k \tag{4.327}$$

as the system

$$\begin{aligned} Ey_1(k) &= y_2(k), \\ Ey_2(k) &= y_3(k), \\ &\;\;\vdots \\ Ey_n(k) + a_1 y_n(k) &+ \cdots + a_{n-1} y_2(k) + a_n y_1(k) = R_k, \end{aligned} \tag{4.328}$$

where $y_1(k) = y_k$. However, we will restrict ourselves to the study of simultaneous linear difference equations in two dependent variables. The method to be presented can be generalized to systems containing a larger number of dependent variables.

Consider the equations

$$\begin{aligned} \phi_1(E)u_k + \psi_1(E)v_k &= F_k, \\ \phi_2(E)u_k + \psi_2(E)v_k &= G_k, \end{aligned} \tag{4.329}$$

where ϕ_1, ϕ_2, ψ_1, ψ_2 are polynomial functions of the operator E; F_k and G_k are given functions of k; and u_k and v_k are the two dependent functions to be determined. If the operator $\phi_2(E)$ is applied to the first of equations (4.329) and the operator $\phi_1(E)$ to the second of equations (4.329), and the two expressions are subtracted, then the following result is obtained:

$$[\phi_2(E)\psi_1(E) - \phi_1(E)\psi_2(E)]v_k = \phi_2(E)F_k - \phi_1(E)G_k. \tag{4.330}$$

Following the same procedure, we also get

$$[\phi_2(E)\psi_1(E) - \phi_1(E)\psi_2(E)]u_k = \psi_1(E)G_k - \psi_2(E)F_k. \tag{4.331}$$

Note that equations (4.330) and (4.331) are linear, inhomogeneous difference equations, for v_k and u_k, respectively. They are both of the same order, with the order n equal to the highest power of E in the operator expression $\phi_2\psi_1 - \phi_1\psi_2$.

Assume that the characteristic equation

$$f(r) = \phi_2(r)\psi_1(r) - \phi_1(r)\psi_2(r) = 0 \tag{4.332}$$

has no multiple roots. (The case of multiple roots causes no problems.) Denote these n roots by $r_1, r_2, \ldots, r_n$. Therefore, equations (4.330) and (4.331) have the general solutions

$$v_k = c_1 r_1{}^k + c_2 r_2{}^k + \cdots + c_n r_n{}^k + V_k, \tag{4.333}$$

$$u_k = \bar{c}_1 r_1{}^k + \bar{c}_2 r_2{}^k + \cdots + \bar{c}_n r_n{}^k + U_k, \tag{4.334}$$

where the c_i and $\bar{c}_i$, $i = 1, 2, \ldots, n$, are two sets of, for the moment, arbitrary constants; V_k is a particular solution of equation (4.330) and U_k is a particular solution of equation (4.331). Note that U_k and V_k are also solutions of the system given by equation (4.329).

Both u_k and v_k contain n arbitrary constants, for a total of $2n$ constants. However, the two sets of constants are not independent. The required relation between them can be found by substituting equations (4.333) and (4.334) into, for example, the first of equations (4.329). We obtain the result

$$\sum_{i=1}^{n} [\bar{c}_i \phi_1(r_i) + c_i \psi_1(r_i)] r_i{}^k = 0. \tag{4.335}$$

Since the roots r_i, $i = 1, 2, \ldots, n$, are distinct, the $r_i{}^k$ are linearly independent functions, and we conclude that

$$\bar{c}_i = -\frac{\psi_1(r_i)}{\phi_1(r_i)} c_i. \tag{4.336}$$

Consequently, the solution to the system given by equation (4.329) is v_k and u_k as expressed in equations (4.333) and (4.334), with the conditions between the coefficients given by equation (4.336). Thus, the solution contains only n arbitrary constants.

4.8.1. Example A

Consider the equations

$$\begin{aligned} (4E - 17)u_k + (E - 4)v_k &= 0, \\ (2E - 1)u_k + (E - 2)v_k &= 0. \end{aligned} \tag{4.337}$$

Solving for v_k gives

$$(E^2 - 8E + 15)v_k = (E-3)(E-5)v_k = 0, \tag{4.338}$$

which has the solution

$$v_k = c_1 3^k + c_2 5^k. \tag{4.339}$$

Using equation (4.336), we obtain

$$\begin{aligned} \bar{c}_1 &= -\frac{3-4}{12-17}c_1 = -\frac{1}{5}c_1, \\ \bar{c}_2 &= -\frac{1}{20-17}c_2 = -\frac{1}{3}c_2. \end{aligned} \tag{4.340a}$$

Therefore, the solution for u_k is

$$u_k = \bar{c}_1 3^k + \bar{c}_2 5^k = -\tfrac{1}{5}c_1 3^k - \tfrac{1}{3}c_2 5^k. \tag{4.340b}$$

4.8.2. Example B

The system

$$\begin{aligned} (E-3)u_k + v_k &= k, \\ 3u_k + (E-5)v_k &= 4^k \end{aligned} \tag{4.341}$$

can be solved for u_k:

$$(E^2 - 8E + 12)u_k = 1 - 4k - 4^k. \tag{4.342}$$

The general solution of this equation is

$$u_k = c_1 2^k + c_2 6^k + 4^{k-1} - \tfrac{4}{5}k - \tfrac{19}{25}. \tag{4.343}$$

From the first of equations (4.341), we obtain

$$\begin{aligned} v_k &= -(E-3)u_k + k \\ &= c_1 2^k - 3c_2 6^k - 4^{k-1} - \tfrac{3}{5}k - \tfrac{34}{25}. \end{aligned} \tag{4.344}$$

Therefore, the general solution of the system given by equations (4.341) is equations (4.343) and (4.344).

4.8.3. Example C

The system of difference equations for u_k and v_k given by

$$\begin{aligned} (2E-3)u_k + 5v_k &= 2, \\ 2u_k + (E-2)v_k &= 7 \end{aligned} \tag{4.345}$$

can be solved for u_k; it satisfies the second-order equation

$$(2E^2 - 7E - 4)u_k = -37, \tag{4.346}$$

and has the general solution

$$u_k = c_1(-\tfrac{1}{2})^k + c_2 4^k + \tfrac{37}{9}. \tag{4.347}$$

Using the first of equations (4.345), we obtain for v_k the result

$$v_k = -\tfrac{1}{5}(2E-3)u_k + \tfrac{2}{5} = \tfrac{11}{9} + \tfrac{4}{5}c_1(-\tfrac{1}{2})^k - c_2 4^k. \tag{4.348}$$

4.8.4. Example D

The system of equations

$$\begin{aligned} (E-1)u_k + 2Ev_k &= 0, \\ -2u_k + (E-1)v_k &= a^k, \end{aligned} \tag{4.349}$$

when solved for v_k gives the equation

$$(E+1)^2 v_k = (a-1)a^k, \tag{4.350}$$

the solution of which is

$$v_k = (c_1 + c_2 k)(-1)^k + \frac{a-1}{(a+1)^2} a^k \tag{4.351}$$

Now, from the second of equations (4.349), we have

$$u_k = \tfrac{1}{2}(E-1)v_k - \tfrac{1}{2}a^k. \tag{4.352}$$

The substitution of v_k from equation (4.351) into the right-hand side of equation (4.351) gives

$$u_k = (-\tfrac{1}{2})[(2c_1 + c_2) + 2c_2 k](-1)^k - \frac{2a^{k+1}}{(a+1)^2}. \tag{4.353}$$

5
LINEAR PARTIAL DIFFERENCE EQUATIONS

5.1. INTRODUCTION

The purpose of this chapter is to provide techniques for obtaining solutions to linear difference equations involving unknown functions of two discrete variables. These equations are called partial difference equations. The functions of interest will be denoted by either of the following representations:

$$z_{k,\ell} \quad \text{or} \quad z(k, \ell). \tag{5.1}$$

In analogy with ordinary difference equations, we introduce the four difference operators

$$\begin{aligned} &E_1 z(k, \ell) = z(k+1, \ell), \qquad E_2 z(k, \ell) = z(k, \ell+1), \\ &\Delta_1 z(k, \ell) = z(k+1, \ell) - z(k, \ell), \\ &\Delta_2 z(k, \ell) = z(k, \ell+1) - z(k, \ell), \end{aligned} \tag{5.2}$$

where

$$\Delta_1 = E_1 - 1 \quad \text{and} \quad \Delta_2 = E_2 - 1. \tag{5.3}$$

It follows immediately that

$$E_1 E_2 z(k, \ell) = E_2 E_1 z(k, \ell) = z(k+1, \ell+1), \tag{5.4}$$

and, consequently,

$$E_1{}^m E_2{}^n z(k, \ell) = z(k+m, \ell+n). \tag{5.5}$$

Let λ and μ be arbitrary constants, and $\phi(E_1, E_2)$ be a polynomial function in E_1 and E_2. An easy calculation shows that

$$E_1{}^m E_2{}^n (\lambda^k \mu^\ell) = \lambda^m \mu^n (\lambda^k \mu^\ell), \tag{5.6}$$

and generally that

$$\phi(E_1, E_2)\lambda^k\mu^\ell = \phi(\lambda, \mu)\lambda^k\mu^\ell. \tag{5.7}$$

A partial difference equation is defined to be a given functional relation among the quantities $z(k, \ell)$, $z(k+1, \ell)$, $z(k, \ell+1)$, $z(k+1, \ell+1)$, $z(k+2, \ell)$, $z(k, \ell+2)$, etc. The following are examples of partial difference equations:

$$z(k+1, \ell) - z(k, \ell+1) = 0, \tag{5.8}$$

$$z(k+2, \ell) + 2z(k+1, \ell+1) + z(k, \ell) = 0, \tag{5.9}$$

$$z(k+1, \ell+1) = 3[z(k, \ell)]^2, \tag{5.10}$$

$$z(k+3, \ell+1) = z(k, \ell) - 5z(k+2, \ell+1)z(k, \ell+1). \tag{5.11}$$

Note that equations (5.8) and (5.9) are linear equations, while equations (5.10) and (5.11) are nonlinear difference equations. Also, equations (5.8) and (5.9) can be conveniently written in the operator forms

$$(E_1 - E_2)z(k, \ell) = 0, \tag{5.8$'$}$$

$$(E_1^2 + 2E_1E_2 + 1)z(k, \ell) = 0. \tag{5.9$'$}$$

We must now define the order of a partial difference equation. If a partial difference equation contains $z(k, \ell)$ and if it also contains terms having arguments $k + m$ and $\ell + n$, where m and n are the largest positive values of m and n, then the equation is said to be of order m with respect to k and order n with respect to ℓ.

For example, equation (5.8) can be rewritten in the form

$$z(k+1, \ell) + z(k, \ell) = z(k, \ell+1) + z(k, \ell), \tag{5.12}$$

and, consequently, is of first order in both k and ℓ. Likewise, equation (5.9) is of second order in k and first order in ℓ, equation (5.10) is first order with respect to k and ℓ, and equation (5.11) is third order in k and first order in ℓ.

Just as for the case of partial differential equations, the explicit solution of a partial difference equation involves certain arbitrary functions of k and ℓ. For linear partial difference equations with constant coefficients, it is straightforward to determine the exact number of arbitrary functions which appear in the general solution. In general, this number is given by the order of either discrete variable k or ℓ. This somewhat paradoxical statement will be made clearer and more precise in the sections to follow.

The major purpose of this chapter is to present a number of techniques for determining the solutions to linear partial difference equations with constant coefficients. The general linear partial difference equation in two variables has the form

$$\phi(E_1, E_2)z(k, \ell) = F(k, \ell), \tag{5.13}$$

where $\phi(E_1, E_2)$ is a polynomial in E_1 and E_2, and $F(k, \ell)$ is a known function of k and ℓ. The general solution of

$$\phi(E_1, E_2)z(k, \ell) = 0 \tag{5.14}$$

is called the homogeneous solution, $z_h(k, \ell)$. Any solution which satisfies equation (5.13) is called a particular solution, $z_p(k, \ell)$. The general solution of equation (5.13) is the sum of its homogeneous solution and any particular solution,

$$z(k, \ell) = z_h(k, \ell) + z_p(k, \ell). \tag{5.15}$$

5.2. SYMBOLIC METHODS

Assume for the homogeneous equation (5.14) that the operator function $\phi(E_1, E_2)$ is factorizable,

$$\phi(E_1, E_2) = \prod_r [E_1 - \psi_r(E_2)]^{s_r}, \tag{5.16}$$

where s_r is a positive integer, $\psi_r(E_2)$ may be an irrational function of E_2, and the sum of the powers s_r is the degree of E_1 in the polynomial function $\phi(E_1, E_2)$. Since the separate factors of $\phi(E_1, E_2)$ are commutative, and, further, since equation (5.14) is a linear equation, the general solution will be the sum of the general solutions of the separate equations

$$[E_1 - \psi_r(E_2)]^{s_r} z(k, \ell) = 0. \tag{5.17}$$

First, consider the simple linear case

$$(E_1 - aE_2 - b)z(k, \ell) = 0, \tag{5.18}$$

where a and b are constants. The corresponding difference equation of first order in k and ℓ is

$$z(k+1, \ell) = az(k, \ell+1) + bz(k, \ell). \tag{5.19}$$

This equation can be written

$$E_1 z(k, \ell) = (aE_2 + b)z(k, \ell). \tag{5.20}$$

Now, E_1 and E_2 act on difference variables. Thus, it is clear that $aE_2 + b$ has no effect on the variable k. Consequently, we may formally solve equation (5.20) to obtain the result

$$z(k, \ell) = (aE_2 + b)^k A(\ell), \tag{5.21}$$

where A is an arbitrary function of the variable ℓ. The operator on the right-hand side of equation (5.21) can be rewritten as

$$\begin{aligned} (aE_2 + b)^k &= b^k \left(1 + \frac{aE_2}{b}\right)^k \\ &= b^k \left(1 + \frac{akE_2}{b} + \frac{a^2 k(k-1)E_2{}^2}{b^2 2!} + \cdots + \left(\frac{a}{b}\right)^k E_2{}^k\right), \end{aligned} \tag{5.22}$$

where in the last step the binomial theorem was used. Since k is a positive integer the expansion, given by equation (5.22), contains only a finite number of terms. Therefore,

$$\begin{aligned} z(k, \ell) = b^k \Bigg(A(\ell) + \frac{ak}{b} A(\ell+1) + \frac{a^2 k(k-1)}{2! b^2} A(\ell+2) \\ + \cdots + \left(\frac{a}{b}\right)^k A(\ell+k)\Bigg) \end{aligned} \tag{5.23}$$

is a general solution to equation (5.18). Note that it contains one arbitrary function of ℓ.

The solution to equation (5.18) can also be obtained by writing equation (5.18) as

$$\left(E_2 - \frac{1}{a}(E_1 - b)\right) z(k, \ell) = 0. \tag{5.24}$$

Therefore, proceeding as before gives

$$\begin{aligned} z(k, \ell) &= \left(-\frac{b}{a}\right)^{\ell} \left(1 - \frac{E_1}{b}\right)^{\ell} B(k) \\ &= \left(-\frac{b}{a}\right)^{\ell} \left(B(k) - \frac{\ell}{b} B(k+1) + \frac{\ell(\ell-1)}{b^2 2!} \right. \\ &\quad \left. \cdot B(k+2) + \cdots + \frac{(-1)^{\ell}}{b^{\ell}} B(k+\ell)\right), \end{aligned} \tag{5.25}$$

where $B(k)$ is an arbitrary function of k.

Next, we consider the case where repeated factors occur. Assume that we have

$$(E_1 - aE_2 - b)^r z(k, \ell) = 0, \tag{5.26}$$

where r is a positive integer. Defining

$$\bar{E} = aE_2 + b, \tag{5.27}$$

and using the fact that

$$(E_1 - \bar{E})^r v(k) = 0 \tag{5.28}$$

has the solution

$$v(k) = \bar{E}^k (A_1 + kA_2 + k^2 A_3 + \cdots + k^{r-1} A_r), \tag{5.29}$$

where the A_i are r arbitrary constants, we obtain the solution to equation (5.27):

$$\begin{aligned} z(k, \ell) &= b^k \left(1 + \frac{aE_2}{b}\right)^k \\ &\quad \cdot [A_1(\ell) + kA_2(\ell) + k^2 A_3(\ell) + \cdots + k^{r-1} A_r(\ell)], \end{aligned} \tag{5.30}$$

where the $A_i(\ell)$ are r arbitrary functions of ℓ. If we use the binomial theorem to expand the operator on the right-hand side of equation (5.30), the general solution of equation (5.26) is obtained in terms of the r arbitrary functions $A_i(\ell)$.

The corresponding solution in terms of r arbitrary functions $B_i(k)$ is

$$z(k, \ell) = \left(-\frac{b}{a}\right)^{\ell}\left(1 - \frac{E_1}{b}\right)^{\ell} \cdot [B_1(k) + \ell B_2(k) + \ell^2 B_3(k) + \cdots + \ell^{r-1} B_r(k)]. \quad (5.31)$$

The following examples illustrate the use of the symbolic method.

5.2.1. Example A

The equation

$$z(k+1, \ell) = 4z(k, \ell+1) \quad (5.32)$$

can be written in symbolic form

$$(E_1 - 4E_2)z(k, \ell) = 0. \quad (5.33)$$

Its solution is

$$z(k, \ell) = (4E_2)^k A(\ell) = 4^k E_2{}^k A(\ell) \quad (5.34)$$

or

$$z(k, \ell) = 4^k A(k + \ell). \quad (5.35)$$

Equation (5.32) is first order in both k and ℓ, and its general solution contains one arbitrary function.

5.2.2. Example B

The equation

$$z(k+2, \ell) = 4z(k, \ell+1) \quad (5.36)$$

can be written

$$(E_1{}^2 - 4E_2)z(k, \ell) = 0. \quad (5.37)$$

We will use two methods to solve this equation.

First, write equation (5.37) in the form

$$(E_2 - \tfrac{1}{4} E_1^2)z(k, \ell) = 0. \tag{5.38}$$

The solution to this equation is

$$z(k, \ell) = (\tfrac{1}{4} E_1^2)^\ell A(k) = 4^{-\ell} A(k + 2\ell). \tag{5.39}$$

Second, we can write equation (5.37) in the factored form

$$(E_1 - 2E_2^{1/2})(E_1 + 2E_2^{1/2})z(k, \ell) = 0. \tag{5.40}$$

Now,

$$(E_1 \pm 2E_2^{1/2})z(k, \ell) = 0 \tag{5.41}$$

has the solution

$$z(k, \ell) = (\mp 2E_2^{1/2})^k \phi_\pm(\ell) = (\mp 2)^k \phi_\pm(\ell + \tfrac{1}{2} k), \tag{5.42}$$

where ϕ_+ and ϕ_- are arbitrary functions of ℓ. Thus, the general solution to equation (5.36) can also be expressed as

$$z(k, \ell) = 2^k[\phi_-(\ell + \tfrac{1}{2} k) + (-)^k \phi_+(\ell + \tfrac{1}{2} k)]. \tag{5.43}$$

Note that equation (5.36) is second order in k and first order in ℓ. This is the reason for obtaining two arbitrary functions in the solution given by equation (5.43) and one in the solution of equation (5.39).

5.2.3. Example C

The partial difference equation of first order in k and ℓ,

$$z(k + 1, \ell + 1) - z(k, \ell + 1) - z(k, \ell) = 0, \tag{5.44}$$

can be written in either of the equivalent forms

$$\begin{aligned} &[E_2(E_1 - 1) - 1]z(k, \ell) = 0, \\ &[E_1 - (1 + E_2^{-1})]z(k, \ell) = 0. \end{aligned} \tag{5.45}$$

Using the latter equation gives

$$z(k, \ell) = (1 + E_2^{-1})^k \phi(\ell), \tag{5.46}$$

where $\phi(\ell)$ is an arbitrary function of ℓ.

From the binomial theorem, we obtain

$$\begin{aligned}(1 + E_2^{-1})^k = 1 + \binom{k}{1} E_2^{-1} + \binom{k}{2} E_2^{-2} \\ + \cdots + \binom{k}{k-1} E_2^{-(k-1)} + E_2^{-k},\end{aligned} \tag{5.47}$$

where

$$\binom{k}{n} = \frac{k!}{n!(k-n)!}, \qquad 0 \le n \le k. \tag{5.48}$$

Putting this result in equation (5.46) and performing the indicated operations gives

$$\begin{aligned}z(k, \ell) = \phi(\ell) + \binom{k}{1} \phi(\ell - 1) + \binom{k}{2} \phi(\ell - 2) \\ + \cdots + \binom{k}{k-1} \phi(\ell - k + 1) + \phi(\ell - k),\end{aligned} \tag{5.49}$$

which contains one arbitrary function.

5.2.4. Example D

The equation

$$z(k + 2, \ell) = z(k, \ell + 1) + z(k, \ell) \tag{5.50}$$

is second order in k and first order in ℓ. In operator form, it can be written as either

$$(E_1^2 - E_2 - 1) z(k, \ell) = 0 \tag{5.51}$$

or

$$(E_1 + \sqrt{1 + E_2})(E_1 - \sqrt{1 + E_2}) z(k, \ell) = 0. \tag{5.52}$$

We will obtain a solution using the first form. Our main reason for not attempting to proceed with the second operator form is that, in general, the operator $\sqrt{1+E_2}$ does not have a well-defined meaning in terms of an infinite-series expansion in E_2.

From equation (5.51), we obtain

$$z(k, \ell) = (-)^{\ell}(1 - E_1^2)^{\ell}\phi(k), \tag{5.53}$$

where $\phi(k)$ is an arbitrary function of k. Now,

$$(1 - E_1^2)^{\ell} = 1 - \binom{\ell}{1} E_1^2 + \binom{\ell}{2} E_1^4 + \cdots + \binom{\ell}{\ell-1} (-)^{\ell-1} E_1^{2(\ell-1)} + (-)^{\ell} E_1^{2\ell}, \tag{5.54}$$

and, therefore,

$$z(k, \ell) = (-)^{\ell} \left[\phi(k) - \binom{\ell}{1} \phi(k+2) + \binom{\ell}{2} \phi(k+4) + \cdots + \binom{\ell}{\ell-1} (-)^{\ell-1} \phi[k + 2(\ell-1)] + (-)^{\ell}\phi(k+2\ell) \right]. \tag{5.55}$$

Again, we point out that equation (5.44) is first order in ℓ and thus the single arbitrary function which appears in equation (5.55).

5.2.5. Example E

The equation

$$z(k+1, \ell) - z(k, \ell+1) = 3z(k, \ell) \tag{5.56}$$

is first order in k and ℓ, and corresponds to the operator equation

$$(E_1 - 2E_2 - 3)z(k, \ell). \tag{5.57}$$

Therefore,

$$\begin{aligned} z(k, \ell) &= (2E_2 + 3)^k \phi(\ell) = 3^k(1 + \tfrac{2}{3}E_2)^k \phi(\ell) \\ &= 3^k[\phi(\ell) + \tfrac{2}{3}k\phi(\ell+1) + \tfrac{2}{9}k(k-1)\phi(\ell+2) + \cdots \\ &\quad + (\tfrac{2}{3})^k \phi(\ell+k)]. \end{aligned} \tag{5.58}$$

5.2.6. Example F

Consider the first-order linear partial difference equation

$$z(k+1, \ell+1) - kz(k, \ell) = 0. \tag{5.59}$$

Note that the coefficient of the $z(k, \ell)$ term is not constant, but is equal to k. In operator form, this equation becomes

$$(E_1 E_2 - k)z(k, \ell) = 0, \tag{5.60}$$

or

$$(E_1 - kE_2^{-1})z(k, \ell) = 0. \tag{5.61}$$

In terms of E_1, this equation is a first-order linear equation whose solution is

$$z(k, \ell) = \prod_{i=1}^{k-1} (iE_2^{-1})\phi(\ell) = (k-1)!E_2^{-k+1}\phi(\ell). \tag{5.62}$$

Therefore, the general solution to equation (5.59) is

$$z(k, \ell) = (k-1)!\phi(\ell - k + 1). \tag{5.63}$$

5.3. LAGRANGE'S AND SEPARATION OF VARIABLES METHODS

A discussion is presented of two methods to obtain special solutions to linear partial difference equations having constant coefficients. In many cases, the two methods give the same results.

First, we consider Lagrange's method. Consider the equation

$$\phi(E_1, E_2)z(k, \ell) = 0, \tag{5.64}$$

where $\phi(E_1, E_2)$ is a polynomial function of E_1 and E_2. Assume that equation (5.64) has a solution of the form

$$z(k, \ell) = \lambda^k \mu^\ell, \tag{5.65}$$

where λ and μ are unspecified constants. Substitution of this result into equation (5.64) shows that this will be a solution provided the following condition is satisfied:

$$\phi(\lambda, \mu) = 0. \tag{5.66}$$

Let $\phi(\lambda, \mu)$ be nth degree in λ and denote the n roots of equation (5.66) as

$$\lambda_i(\mu), \qquad 1 \le i \le n. \tag{5.67}$$

Therefore, according to equations (5.65) and (5.66), the following are solutions to equation (5.64):

$$z_i(k, \ell) = [\lambda_i(\mu)]^k \mu^\ell, \qquad 0 \le i \le n. \tag{5.68}$$

Since equation (5.64) is linear, the sum of all such expressions for all possible values of μ will also be a solution. If we let $C_i(\mu)$ and $D_i(\mu)$ be arbitrary functions of μ, then

$$z(k, \ell) = \sum_{i=1}^{n} z_i(k, \ell) \tag{5.69}$$

is a solution of equation (5.64), where for discrete values of μ

$$z_i(k, \ell) = \sum_{\mu} C_i(\mu)[\lambda_i(\mu)]^k \mu^\ell, \tag{5.70}$$

and for continuous values of μ

$$z_i(k, \ell) = \int_{-\infty}^{\infty} D_i(\mu)[\lambda_i(\mu)]^k \mu^\ell \, d\mu. \tag{5.71}$$

The separation of variables method can often be used to obtain special solutions to linear partial difference equations having the form

$$\psi(E_1, E_2, k, \ell)z(k, \ell) = 0, \tag{5.72}$$

where ψ is a polynomial function of E_1 and E_2. The basic principle of the method is to assume that $z(k, \ell)$ has the structure

$$z(k, \ell) = C_k D_\ell. \tag{5.73}$$

Further, assume that when equation (5.73) is substituted into equation (5.72), an equation of the type

$$\frac{f_1(E_1, k)C_k}{f_2(E_1, k)C_k} = \frac{g_1(E_2, \ell)D_\ell}{g_2(E_2, \ell)D_\ell} \tag{5.74}$$

is obtained. Under these conditions, C_k and D_ℓ satisfy the following ordinary difference equations:

$$\begin{aligned} f_1(E_1, k)C_k &= \alpha f_2(E_1, k)C_k, \\ g_1(E_2, \ell)D_\ell &= \alpha g_2(E_2, \ell)D_\ell, \end{aligned} \tag{5.75}$$

where α is an arbitrary constant.

The basic argument used to obtain equations (5.75) relies on the fact that if equations (5.74) hold, then it is easily seen that the left-handside is a function only of k, while the right-hand side is a function only of ℓ. The only possibility that can hold is for left- and right-hand sides to be equal to an arbitrary constant since k and ℓ are independent variables. Once we have obtained solutions for C_k and D_ℓ, summing over α gives additional solutions just as was the case for Lagrange's method. In general, we expect that Lagrange's method will provide special solutions for any partial difference equation having the form given by equation (5.64). This may not be true for the separation of variables method.

We will now work a number of examples using both methods where suitable.

5.3.1. Example A

The substitution of equation (5.65) into the equation

$$z(k+1, \ell) = z(k, \ell+1) \tag{5.76}$$

gives

$$\lambda^{k+1}\mu^{\ell} = \lambda^{k}\mu^{\ell+1} \quad \text{or} \quad \lambda = \mu. \tag{5.77}$$

Therefore, a special solution of equation (5.76) is

$$z(k, \ell) = \mu^{k+\ell}. \tag{5.78}$$

Summing solutions of this form gives

$$z(k, \ell) = \int_{-\infty}^{\infty} C(\mu)\mu^{k+\ell}\, d\mu, \tag{5.79}$$

where $C(\mu)$ is chosen so that the integral is defined. From equation (5.79) we conclude that a general solution of equation (5.76) is

$$z(k, \ell) = f(k + \ell), \tag{5.80}$$

where f is an arbitrary function of $k + \ell$. This result has already been obtained by the method of Section 5.2.

Using the separation of variables method, where $z(k, \ell) = C_k D_\ell$, we rewrite equation (5.76) as

$$C_{k+1} D_\ell = C_k D_{\ell+1}, \tag{5.81}$$

or

$$\frac{C_{k+1}}{C_k} = \frac{D_{\ell+1}}{D_\ell} = \alpha = \text{arbitrary constant.} \tag{5.82}$$

These latter equations have the solutions

$$C_k = A\alpha^k \quad \text{and} \quad D_\ell = B\alpha^\ell, \tag{5.83}$$

where A and B are arbitrary constants. If we let $\alpha = \mu$, then we obtain the same situation as in the above calculation using Lagrange's method. Thus, the solution is given by equation (5.80).

5.3.2. Example B

The equation

$$z(k + 1, \ell) = 4z(k, \ell + 1) \tag{5.84}$$

leads to the conclusion that $\lambda = 4\mu$. Therefore, equation (5.84) has the special solutions

$$z(k, \ell) = 4^k \mu^{k+\ell}. \tag{5.85}$$

Summing expressions of this form allows the conclusion

$$z(k, \ell) = 4^k f(k + \ell), \tag{5.86}$$

where f is an arbitrary function of $k + \ell$.

The separation of variables method, where $z_p(k, \ell) = C_k D_\ell$, gives

$$C_{k+1} D_\ell = 4C_k D_{\ell+1}, \tag{5.87}$$

which can be written in either of the forms

$$\frac{C_{k+1}}{4C_k} = \frac{D_{\ell+1}}{D_\ell} = \alpha \quad \text{or} \quad \frac{C_{k+1}}{C_k} = \frac{4D_{\ell+1}}{D_\ell} = \alpha. \tag{5.88}$$

The respective solutions to these equations give

$$z(k, \ell) = 4^k \alpha^{k+\ell} \quad \text{or} \quad z(k, \ell) = 4^{-\ell} \alpha^{k+\ell}. \tag{5.89}$$

Summing over α allows us to obtain the general solution

$$z(k, \ell) = 4^k f(k+\ell) \quad \text{or} \quad z(k, \ell) = 4^{-\ell} g(k+\ell), \tag{5.90}$$

where f and g are arbitrary functions of $k + \ell$.

5.3.3. Example C

Let $p + q = 1$ in the equation

$$z(k, \ell) = pz(k+1, \ell-1) + qz(k-1, \ell+1). \tag{5.91}$$

Lagrange's method requires that

$$p\lambda^2 - \mu\lambda + q^2\mu^2 = 0. \tag{5.92}$$

The solutions, $\lambda_1(\mu)$ and $\lambda_2(\mu)$, of this equation are

$$\lambda_1(\mu) = \mu, \qquad \lambda_2(\mu) = \mu q/p. \tag{5.93}$$

This gives the following two particular solutions:

$$z_1(k, \ell) = \mu^{k+\ell} \quad \text{and} \quad z_2(k, \ell) = (q/p)^k \mu^{k+\ell}. \tag{5.94}$$

Summing these expressions and adding the results gives the general solution

$$z(k, \ell) = g(k+\ell) + (q/p)^k h(k+\ell). \tag{5.95}$$

Note that the separation of variables method does not work for this equation, since if $z(k, \ell) = C_k D_\ell$, the following result is obtained:

$$C_k D_\ell = pC_{k+1} D_{\ell-1} + qC_{k-1} D_{\ell+1}, \tag{5.96}$$

and the necessary separation into expressions that depend only on k and ℓ cannot be done.

5.3.4. Example D

Lagrange's method cannot be applied to the equation

$$z(k, \ell + 1) = z(k - 1, \ell) + kz(k, \ell), \tag{5.97}$$

since one of the coefficients depends on k. However, if we let $z(k, \ell) = C_k D_\ell$, then

$$C_k D_{\ell+1} = C_{k-1} D_\ell + kC_k D_\ell, \tag{5.98}$$

which can be rewritten as

$$\frac{D_{\ell+1}}{D_\ell} = \frac{C_{k-1} + kC_k}{C_k} = \alpha, \tag{5.99}$$

where α is an arbitrary constant. Therefore, D_ℓ and C_k satisfy the first-order difference equations

$$D_{\ell+1} = \alpha D_\ell, \qquad (\alpha - k)C_k = C_{k-1}, \tag{5.100}$$

the solutions of which are, respectively,

$$D_\ell = A\alpha^\ell, \qquad C_k = \frac{B(-1)^k}{\Gamma(k - (\alpha - 1)]}. \tag{5.101}$$

Summing over α gives

$$z(k, \ell) = (-)^k \sum_\alpha \frac{\phi(\alpha)\alpha^\ell}{\Gamma[k - (\alpha - 1)]}, \tag{5.102}$$

where ϕ is an arbitrary function of α.

5.3.5. Example E

Applying the method of Lagrange to the partial difference equation

$$z(k + 1, \ell) - 2z(k, \ell + 1) - 3z(k, \ell) = 0 \tag{5.103}$$

gives $\lambda = 2\mu + 3$, and special solutions having the form

$$z_p(k, \ell) = (2\mu + 3)^k \mu^\ell. \tag{5.104}$$

General solutions can be obtained by summing over μ; doing this gives

$$z(k, \ell) = \sum_{\mu} C(\mu)(2\mu + 3)^k \mu^\ell, \tag{5.105}$$

where $C(\mu)$ is an arbitrary function of μ. Now,

$$\begin{aligned}(2\mu + 3)^k &= 3^k \left(1 + \frac{2\mu}{3}\right)^k \\ &= 3^k \left[1 + k\frac{2\mu}{3} + \frac{k(k-1)}{2!}\left(\frac{2\mu}{3}\right)^2 \right. \\ &\qquad \left. + \cdots + k\left(\frac{2\mu}{3}\right)^{k-1} + \left(\frac{2\mu}{3}\right)^k\right],\end{aligned} \tag{5.106}$$

and, therefore, equation (5.105) can be written

$$\begin{aligned}z(k, \ell) = 3^k[H(\ell) + \tfrac{2}{3}kH(\ell + 1) + \tfrac{2}{9}k(k-1)H(\ell + 2) \\ + \cdots + (\tfrac{2}{3})^k H(\ell + k)],\end{aligned} \tag{5.107}$$

where $H(\ell)$ is an arbitrary function of ℓ and we have used the results

$$\begin{aligned}H(\ell) &= \sum_{\alpha} C(\mu)\mu^\ell, \\ H(\ell + m) &= \sum_{\mu} C(\mu)\mu^{\ell + m}.\end{aligned} \tag{5.108}$$

Applying the method of separation of variables gives for C_k and D_ℓ the equations

$$\frac{C_{k+1}}{C_k} = \frac{2D_{\ell+1} + 3D_\ell}{D_\ell} = \alpha, \tag{5.109}$$

the solutions of which are

$$C_k = A\alpha^k, \qquad D_\ell = B\left(\frac{\alpha - 3}{2}\right)^\ell, \tag{5.110}$$

where A and B are arbitrary constants. If we let

$$\mu = \frac{\alpha - 3}{2}, \tag{5.111}$$

then $z(k, \ell) = C_k D_\ell$ becomes

$$z(k, \ell) = AB(3 + 2\mu)^k \mu^\ell, \tag{5.112}$$

which is the same as equation (5.104). Therefore, the separation of variables method gives the same solution as presented in equation (5.107).

5.3.6. Example F

The Lagrange method applied to the equation

$$z(k + 2, \ell) = 4z(k, \ell + 1) \tag{5.113}$$

gives $\lambda^2 = 4\mu$ or $\lambda_1 = 2\lambda^{1/2}$ and $\lambda_2 = -2\mu^{1/2}$. Therefore, special solutions are

$$z_1(k, \ell) = (2\mu^{1/2})^k \mu^\ell \tag{5.114}$$

and

$$z_2(k, \ell) = (-2\mu^{1/2})^k \mu^\ell. \tag{5.115}$$

Multiplying these two equations by $C_1(\mu)$ and $C_2(\mu)$ and summing over μ gives the general solution

$$z(k, \ell) = 2^k[f(k + 2\ell) + (-1)^k g(k + 2\ell)], \tag{5.116}$$

where f and g are arbitrary functions of $k + 2\ell$.

Likewise, the separation of variables method, $z_p(k, \ell) = C_k D_\ell$, gives the equations

$$\frac{C_{k+2}}{4C_k} = \frac{D_{\ell+1}}{D_\ell} = \alpha^2, \tag{5.117}$$

where we have written the "separation constant" in the form α^2. The solutions to equations (5.117) allow us to determine $z_p(k, \ell)$; it is

$$z_p(k, \ell) = [A_1\alpha^{k+2\ell} + A_2(-)^k\alpha^{k+2\ell}]2^k, \tag{5.118}$$

where A_1 and A_2 are arbitrary constants. Summing over α gives the general solution expressed by equation (5.116).

5.3.7. Example C

The equation

$$z(k+3, \ell) - 3z(k+2, \ell+1) + 3z(k+1, \ell+2) - z(k, m+3) = 0 \tag{5.119}$$

cannot be solved by the method of separation of variables. However, the Lagrange's method does apply and gives

$$\lambda^3 - 3\lambda^2\mu + 3\lambda\mu^2 - \mu^3 = 0, \tag{5.120}$$

or

$$(\lambda - \mu)^3 = 0. \tag{5.121}$$

Therefore, $\lambda = \mu$ is a triple root and linearly independent particular solutions are

$$\mu^{k+\ell}, \qquad k\mu^{k+\ell}, \qquad \text{and} \quad k^2\mu^{k+\ell}.$$

Multiplying each of these expressions by an arbitrary function of μ and summing gives the general solution

$$z(k, \ell) = f(k+\ell) + kg(k+\ell) + k^2h(k+\ell), \tag{5.122}$$

where f, g, and h are arbitrary functions of $k + \ell$.

Finally, we will solve equation (5.119) using the symbolic method of Section 5.2. To proceed, note that equation (5.119) can be written

$$(E_1^3 - 3E_2E_1^2 + 3E_2^2E_1 - E_2^3)z(k, \ell) = 0 \tag{5.123}$$

or

$$(E_1 - E_2)^3z(k, \ell) = 0. \tag{5.124}$$

Now equation (5.124) is of the same form as equation (5.26) with $a = 1$ and $b = 0$. Therefore, using equation (5.30), which for this case becomes

$$z(k, \ell) = E_2{}^k[A_1(\ell) + kA_2(\ell) + k^2A_3(\ell)], \tag{5.125}$$

we obtain the general solution

$$z(k, \ell) = A_1(\ell + k) + kA_2(\ell + k) + k^2A_3(\ell + k), \tag{5.126}$$

which is the same as equation (5.122).

5.4. LAPLACE'S METHOD

This method can be used when the sum or difference of the arguments of all the $z(k + r, \ell + s)$ that appear in a partial difference equation are constant. For example, consider the following difference equation with constant coefficients:

$$A_0z(k, \ell) + A_1z(k - 1, \ell - 1) + \cdots + A_nz(k - n, \ell - n) = W(k, \ell), \tag{5.127}$$

where $W(k, \ell)$ is a given function. Note that the difference of the arguments, $k - \ell$, is equal to a constant. Now, if we set $k - \ell = m$ and put

$$z(k, \ell) = z(k, k - m) = v_k, \tag{5.128a}$$

then equation (5.127) becomes

$$A_0v_k + A_1v_{k-1} + \cdots + A_nv_{k-n} = W(k, k - m). \tag{5.128b}$$

The solution of the latter equation can be expressed as

$$v_k = C_1\phi_1(k) + C_2\phi_2(k) + \cdots + C_n\phi_n(k) + P(k, m), \tag{5.129}$$

where the C_i are arbitrary constants; $\phi_i(k)$, for $i = 1$ to n, is a set of linearly independent solutions; and $P(k, m)$ is a particular solution. Since an arbitrary constant can be considered a function of another constant, we can recover the solution to our original partial difference equation by replacing m in equations (5.129) by $k - \ell$ and the arbitrary constants by arbitrary functions of $k - \ell$. Doing this gives

$$\begin{aligned} z(k, \ell) = {} & f_1(k - \ell)\phi_1(k) + f_2(k - \ell)\phi_2(k) + \cdots \\ & + f_n(k - \ell)\phi_n(k) + P(k, k - \ell). \end{aligned} \tag{5.130}$$

The above method can also be applied to partial difference equations having coefficients which are functions of k and ℓ.

The following examples illustrate the use of Laplace's method.

5.4.1. Example A

The equation

$$z(k, \ell) = pz(k+1, \ell-1) + (1-p)z(k-1, \ell+1) \tag{5.131}$$

has the sum of its various arguments equal to a constant. Let $k + \ell = m$ and define

$$v_k = z(k, m-k). \tag{5.132}$$

Therefore, equation (5.131) becomes

$$v_k = pv_{k+1} + (1-p)v_{k-1}, \tag{5.133}$$

which is a second-order ordinary difference equation whose solution is

$$v_k = C_1 + C_2\left(\frac{1-p}{p}\right)^k. \tag{5.134}$$

Replacing C_1 and C_2 by arbitrary functions of $k + \ell$ gives the general solution to equation (5.131),

$$z(k, \ell) = f_1(k+\ell) + \left(\frac{1-p}{p}\right)^k f_2(k+\ell). \tag{5.135}$$

5.4.2. Example B

Consider the equation

$$z(k, \ell) + 2z(k-1, \ell-1) = \ell, \tag{5.136}$$

and let $k - \ell = m$. Therefore, setting

$$v_k = z(k, k-m) \tag{5.137}$$

gives

$$v_k + 2v_{k-1} = k - m. \tag{5.138}$$

The solution to equation (5.138) is

$$v_k = C_1(-2)^k + (2/9 - 1/3 m) + 1/3 k. \tag{5.139}$$

If we now replace m by $k - \ell$ and C_1 by $f(k - \ell)$, we obtain the general solution to equation (5.136),

$$z(k, \ell) = (-2)^k f(k - \ell) + 2/9 + 1/3 \ell. \tag{5.140}$$

5.4.3. Example C

The two previous examples each had constant coefficients. Now consider the equation

$$z(k, \ell) - kz(k - 1, \ell - 1) = 0, \tag{5.141}$$

where the difference of the arguments is a constant. Letting $k - \ell = m$ and setting

$$v_k = z(k, k - m) \tag{5.142}$$

gives on substitution into equation (5.141) the result

$$v_k - kv_{k-1} = 0. \tag{5.143}$$

The solution to the latter equation is

$$v_k = Ck!, \tag{5.144}$$

where C is an arbitrary constant. Replacing C by an arbitrary function of $k - \ell$ gives the general solution of equation (5.141),

$$z(k, \ell) = f(k - \ell)k!. \tag{5.145}$$

5.4.4. Example D

The equation

$$z(k + 3, \ell) - 3z(k + 2, \ell + 1) + 3z(k + 1, \ell + 2) - z(k, \ell + 3) = 0 \tag{5.146}$$

has the sum of its arguments equal to a constant. Therefore, setting

$$k + \ell = m, \qquad v_k = z(k, m - k), \tag{5.147}$$

we obtain

$$v_{k+3} - 3v_{k+2} + 3v_{k+1} - v_k = 0, \tag{5.148}$$

the characteristic equation of which is

$$r^3 - 3r^2 + 3r - 1 = (r-1)^3 = 0. \tag{5.149}$$

Therefore, its solution is

$$v_k = C_1 + C_2 k + C_3 k^2, \tag{5.150}$$

and we conclude that the general solution of equation (5.146) is

$$z(k, \ell) = f_1(k+\ell) + k f_2(k+\ell) + k^2 f_3(k+\ell). \tag{5.151}$$

5.5. PARTICULAR SOLUTIONS

Thus far in this chapter, we have discussed methods for obtaining solutions to linear, homogeneous partial difference equations. We now consider several procedures for obtaining particular solutions to linear, inhomogeneous partial difference equations with constant coefficients.

First, consider equations of the form

$$\phi(E_1, E_2) z(k, \ell) = F(k, \ell), \tag{5.152}$$

where $\phi(E_1, E_2)$ is a polynomial function of its arguments and $F(k, \ell)$ is a given polynomial function of k and ℓ. If we use

$$E_1 = 1 + \Delta_1, \qquad E_2 = 1 + \Delta_2, \tag{5.153}$$

then the operator on the left-hand side of equation (5.152) becomes

$$\phi(E_1, E_2) = \phi(1+\Delta_1, 1+\Delta_2) \tag{5.154}$$

and we obtain the particular solution $z_p(k, \ell)$, which can be written as

$$z_p(k, \ell) = \frac{1}{\phi(1+\Delta_1, 1+\Delta_2)} F(k, \ell). \tag{5.155}$$

Expanding the operator $1/\phi(1+\Delta_1, 1+\Delta_2)$ in powers of Δ_1 and Δ_2 gives

$$\frac{1}{\phi(1+\Delta_1, 1+\Delta_2)} = \phi_0 + \phi_1\Delta_1 + \phi_2\Delta_2 + \phi_{11}\Delta_1^2 + \phi_{12}\Delta_1\Delta_2 + \phi_{22}\Delta_2^2 + \cdots, \quad (5.156)$$

where ϕ_0, ϕ_1, etc., are known constants. Therefore, the particular solution is

$$z_p(k, \ell) = (\phi_0 + \phi_1\Delta_1 + \phi_2\Delta_2 + \cdots)F(k, \ell). \quad (5.157)$$

Since $F(k, \ell)$ is a polynomial function of k and ℓ, the right-hand side of equation (5.157) will only contain a finite number of terms after all the various operations have been performed.

Now consider the case where the equation of interest is

$$\phi(E_1, E_2)z(k, \ell) = \lambda^{ak+b\ell}F(k, \ell), \quad (5.158)$$

where $\phi(E_1, E_2)$ is a polynomial function of E_1 and E_2; λ, a, and b are constants; and $F(k, \ell)$ is a polynomial function of k and ℓ. Let the solution be written

$$z(k, \ell) = \lambda^{ak+b\ell}w(k, \ell). \quad (5.159)$$

An easy calculation shows that

$$E_1^m E_2^n[\lambda^{ak+b\ell}w(k, \ell)] = \lambda^{am+bn}\lambda^{ak+b\ell}E_1^m E_2^n w(k, \ell), \quad (5.160)$$

and, consequently, we have the result

$$\phi(E_1, E_2)[\lambda^{ak+b\ell}w(k, \ell)] = \lambda^{ak+b\ell}\phi(\lambda^a E_1, \lambda^b E_2)w(k, \ell). \quad (5.161)$$

Therefore, equation (5.158) becomes under this transformation

$$\phi(\lambda^a E_1, \lambda^b E_2)w(k, \ell) = F(k, \ell), \quad (5.162)$$

which is of the form considered above.

The following examples illustrate the use of the above methods and indicate how other special cases may be handled.

5.5.1. Example A

The equation

$$z(k+1, \ell) - 4z(k, \ell+1) = 6k^2\ell + 4 \quad (5.163)$$

can be written as

$$(E_1 - 4E_2)z(k, \ell) = 6k^2\ell + 4. \tag{5.164}$$

The homogeneous equation

$$(E_1 - 4E_2)z_h(k, \ell) = 0 \tag{5.165}$$

has the solution

$$z_h(k, \ell) = 4^k f(k + \ell), \tag{5.166}$$

where f is an arbitrary function of $k + \ell$.

The particular solution is, from equation (5.155), given by the expression

$$\begin{aligned} z_p(k, \ell) &= \frac{1}{E_1 - 4E_2}(6k^2\ell + 4) \\ &= \frac{-1}{3[1 - \frac{1}{3}(\Delta_1 - 4\Delta_2)]}(6k^2\ell + 4). \end{aligned} \tag{5.167}$$

Expanding the operator on the right-hand side of equation (5.167) gives

$$\begin{aligned} &\left(-\frac{1}{3}\right)\left[1 + \frac{\Delta_1 - 4\Delta_2}{3} + \left(\frac{\Delta_1 - 4\Delta_2}{3}\right)^2 + \cdots\right](6k^2\ell + 4) \\ &\qquad = -2k^2\ell - \frac{4k\ell}{3} + \frac{8k^2}{3} + \frac{32k}{9} - \frac{10\ell}{9} + \frac{20}{9}, \end{aligned} \tag{5.168}$$

which is the required particular solution $z_p(k, \ell)$.

Therefore, equation (5.163) has the solution

$$z(k, \ell) = z_h(k, \ell) + z_p(k, \ell), \tag{5.169}$$

where $z_h(k, \ell)$ and $z_p(k, \ell)$ are given, respectively, by equations (5.166) and (5.168).

5.5.2. Example B

Consider the equation

$$z(k + 1, \ell + 1) + 2z(k, \ell) = k^2 + \ell + 1, \tag{5.170}$$

or

$$(E_1E_2 + 2)z(k, \ell) = k^2 + \ell + 1. \tag{5.171}$$

The particular solution is

$$\begin{aligned} z_p(k, \ell) &= \frac{1}{E_1E_2 + 2}(k^2 + \ell + 1) \\ &= \frac{1}{3}\frac{1}{1 + \frac{1}{3}(\Delta_1 + \Delta_2 + \Delta_1\Delta_2)}(k^2 + \ell + 1) \\ &= \tfrac{1}{3}[1 - \tfrac{1}{3}(\Delta_1 + \Delta_2 + \Delta_1\Delta_2) \\ &\quad + \tfrac{1}{9}(\Delta_1^2 + \Delta_2^2 + \Delta_1^2\Delta_2^2 + 2\Delta_1\Delta_2 + 2\Delta_1^2\Delta_2 \\ &\quad + 2\Delta_1\Delta_2^2) + \cdots](k^2 + \ell + 1) \\ &= \tfrac{1}{27}(9k^2 - 6k + 9\ell + 5). \end{aligned} \tag{5.172}$$

The homogeneous equation

$$(E_1E_2 + 2)z_h(k, \ell) = 0 \tag{5.173}$$

has the solution

$$z_h(k, \ell) = (-2)^k f(\ell - k). \tag{5.174}$$

Therefore, the general solution of equation (5.170) is

$$\begin{aligned} z(k, \ell) &= z_h(k, \ell) + z_p(k, \ell) \\ &= (-2)^k f(\ell - k) + \tfrac{1}{27}(9k^2 - 6k + 9\ell + 5). \end{aligned} \tag{5.175}$$

5.5.3. Example C

The equation

$$z(k + 1, \ell + 1) + 2z(k, \ell) = 3^k(k^2 + \ell + 1) \tag{5.176}$$

corresponds to equation (5.158) with the following identifications: $\phi(E_1, E_2) = E_1E_2 + 2$, $\lambda = 3$, $a = 1$, $b = 0$, and $F(k, \ell) = k^2 + \ell + 1$. From equation (5.159) we see that the solution $z(k, \ell)$ can be expressed as

$$z(k, \ell) = 3^k w(k, \ell), \tag{5.177}$$

where $w(k, \ell)$ satisfies the equation

$$(3E_1E_2 + 2)w(k, \ell) = k^2 + \ell + 1. \tag{5.178}$$

The particular solution to this equation can be obtained as follows:

$$\begin{aligned} w_p(k, \ell) &= \frac{1}{2 + 3E_1E_2}(k^2 + \ell + 1) \\ &= \frac{1}{5}\frac{1}{1 + 3/5(\Delta_1 + \Delta_2 + \Delta_1\Delta_2)}(k^2 + \ell + 1) \\ &= 1/5[1 - 3/5(\Delta_1 + \Delta_2 + \Delta_1\Delta_2) \\ &\quad + 9/25(\Delta_1{}^2 + \Delta_2{}^2 + \Delta_1{}^2\Delta_2{}^2 + \cdots) + \cdots](k^2 + \ell + 1) \\ &= 1/125(25k^2 - 30k + 25\ell + 13). \end{aligned} \tag{5.179}$$

Likewise, the homogeneous equation

$$(3E_1E_2 + 2)w_h(k, \ell) = 0 \tag{5.180}$$

has the solution

$$w_h(k, \ell) = (-2/3)^k f(\ell - k), \tag{5.181}$$

where f is an arbitrary function of $\ell - k$. Therefore, the complete solution of equation (5.178) is given by the sum of the expressions in equations (5.179) and (5.181). Finally, putting these results into equation (5.177) gives

$$z(k, \ell) = (-2)^k f(\ell - k) + \frac{3^k}{125}(25k^2 - 30k + 25\ell + 13). \tag{5.182}$$

5.5.4. Example D

Let us investigate the case where the inhomogeneous term contains a factor which also appears in the solution of the homogeneous equation. Consider the first-order equation

$$(aE_1 + bE_2 + c)z(k, \ell) = F(k, \ell), \tag{5.183}$$

where a, b, and c are constants, and $F(k, \ell)$ satisfies the condition

$$(aE_1 + bE_2 + c)F(k, \ell) = 0. \tag{5.184}$$

If we assume that the particular solution has the form

$$z_p(k, \ell) = (Ak + B\ell)F(k, \ell), \tag{5.185}$$

where A and B are unknown constants, then substitution of this result into equation (5.183) gives

$$(aAE_1 + bBE_2 - 1)F(k, \ell) = 0. \tag{5.186}$$

Comparison of equations (5.184) and (5.186) allows the conclusion

$$A = B = -1/c. \tag{5.187}$$

Therefore, for this case, the particular solution

$$z_p(k, \ell) = (-1/c)(k + \ell)F(k, \ell). \tag{5.188}$$

5.5.5. Example E

Assume that $c = 0$ in equation (5.183) and continue to assume that $F(k, \ell)$ is a solution to the homogeneous equation. Under these conditions, we have

$$(aE_1 + bE_2)z(k, \ell) = F(k, \ell) \tag{5.189}$$

and

$$(aE_1 + bE_2)F(k, \ell) = 0. \tag{5.190}$$

The solution to the last equation is

$$F(k, \ell) = (-b/a)^k f(\ell + k), \tag{5.191}$$

where f is an arbitrary function of $\ell + k$.

Examination of the left-hand side of equation (5.189) shows that it is of a form such that Laplace's method can be used to obtain a solution. If we let

$$k + \ell = m = \text{constant}, \qquad v_k = z(k, \ell) = z(k, m - k), \tag{5.192}$$

then v_k satisfies the first-order inhomogeneous equation

$$av_{k+1} + bv_k = (-b/a)^k f(m), \tag{5.193}$$

where we have used the results of equations (5.191) and (5.192) to replace the right-hand side of equation (5.189). Note that $f(m)$ is a constant. Solving for v_k gives

$$v_k = A\left(-\frac{b}{a}\right)^k - \frac{1}{b} f(m) k \left(-\frac{b}{a}\right)^k, \tag{5.194}$$

where A is an arbitrary constant. Replacing m by $\ell + k$ and A by an arbitrary function of $\ell + k$ gives the complete solution to equation (5.189), under the assumption of equation (5.190),

$$z(k, \ell) = \left(-\frac{b}{a}\right)^k g(\ell + k) - \frac{k}{b}\left(-\frac{b}{a}\right)^k f(\ell + k). \tag{5.195}$$

5.5.6. Example F

An example of another special case is the equation

$$z(k+1, \ell+1) - z(k, \ell) = F(k - \ell), \tag{5.196}$$

for which the function on the right-hand side is a solution to the homogeneous part of the equation. This equation is of a form such that Laplace's method can be applied. Applying this technique gives the result

$$z(k, \ell) = f(k - \ell) + \tfrac{1}{2}(k + \ell) F(k - \ell), \tag{5.197}$$

where f is an arbitrary function of $k - \ell$.

Similarly, consider the equation

$$z(k+1, \ell+1) - z(k, \ell) = \lambda^k F(k - \ell) w(k, \ell), \tag{5.198}$$

where λ is a constant and $w(k, \ell)$ is a given polynomial function of k and ℓ. Letting

$$k - \ell = m = \text{constant}, \qquad v_k = z(k, k - m), \tag{5.199}$$

gives the following first-order difference equation:

$$v_{k+1} - v_k = \lambda^k F(m) w(k, k - m), \tag{5.200}$$

which can easily be solved. Equation (5.199) can then be used to obtain $z(k, \ell)$.

5.6. SIMULTANEOUS EQUATIONS WITH CONSTANT COEFFICIENTS

Consider two functions of k and ℓ: $u(k, \ell)$ and $v(k, \ell)$. Let $\phi_1(E_1, E_2)$, $\phi_2(E_1, E_2)$, $\psi_1(E_1, E_2)$, and $\psi_2(E_1, E_2)$ be polynomial functions of the operators E_1 and E_2. Further, let $F(k, \ell)$ and $G(k, \ell)$ be given functions of k and ℓ. The relationships

$$\begin{aligned} \phi_1(E_1, E_2)u(k, \ell) + \phi_2(E_1, E_2)v(k, \ell) &= F(k, \ell), \\ \psi_1(E_1, E_2)u(k, \ell) + \psi_2(E_1, E_2)v(k, \ell) &= G(k, \ell) \end{aligned} \tag{5.201}$$

define a pair of simultaneous linear ordinary difference equations for $u(k, \ell)$ and $v(k, \ell)$.

These equations can be solved by first eliminating $v(k, \ell)$ and solving the resulting equation for $u(k, \ell)$. Likewise, $u(k, \ell)$ can be eliminated and the equation for $v(k, \ell)$ solved. These solutions will contain a number of arbitrary functions. However, substitution of $u(k, \ell)$ and $v(k, \ell)$ back into equation (5.201) will allow the determination of the proper number of arbitrary functions which should be present in the final solution.

5.6.1. Example A

The simultaneous equations

$$\begin{aligned} E_1u(k, \ell) + E_2v(k, \ell) &= 0, \\ E_2u(k, \ell) + E_1v(k, \ell) &= 1 \end{aligned} \tag{5.202}$$

can be solved in turn for $u(k, \ell)$ and $v(k, \ell)$ to give

$$(E_1^2 - E_2^2)u(k, \ell) = -1 \tag{5.203}$$

and

$$(E_1^2 - E_2^2)v(k, \ell) = 1. \tag{5.204}$$

The general solutions to equations (5.203) and (5.204) are, respectively,

$$u(k, \ell) = f(k + \ell) + (-1)^k g(k + \ell) - \tfrac{1}{2}k \tag{5.205}$$

and

$$v(k, \ell) = F(k + \ell) + (-1)^k G(k + \ell) + \tfrac{1}{2}k, \tag{5.206}$$

where f, g, F, and G are arbitrary functions of $k + \ell$. The substitution of equations (5.205) and (5.206) into the first of equations (5.202) gives

$$E_1u(k, \ell) + E_2v(k, \ell) = f(k + \ell + 1) - (-1)^k g(k + \ell + 1) - \tfrac{1}{2}k - \tfrac{1}{2} + F(k + \ell + 1) + (-1)^k G(k + \ell + 1) + \tfrac{1}{2}k, \quad (5.207)$$

and we conclude that

$$F(k + \ell) = \tfrac{1}{2} - f(k + \ell), \qquad G(k + \ell) = g(k + \ell). \quad (5.208)$$

The second of equations (5.202) is also satisfied by these relations.

We conclude that the general solution of the simultaneous pair of difference equations given by equation (5.202) is

$$\begin{aligned} u(k, \ell) &= f(k + \ell) + (-)^k g(k + \ell) - \tfrac{1}{2}k, \\ v(k, \ell) &= \tfrac{1}{2} - f(k + \ell) + (-1)^k g(k + \ell) + \tfrac{1}{2}k. \end{aligned} \quad (5.209)$$

5.6.2. Example B

Consider the equations

$$\begin{aligned} 2(2E_1E_2 - 1)u(k, \ell) - (3E_1E_2 - 1)v(k, \ell) &= 1, \\ 2(E_1E_2 - 1)u(k, \ell) + (E_1E_2 - 1)v(k, \ell) &= k. \end{aligned} \quad (5.210)$$

Eliminating $v(k, \ell)$ gives

$$(E_1E_2 - 1)(5E_1E_2 - 2)u(k, \ell) = k + \tfrac{3}{2}, \quad (5.211)$$

the general solution of which is

$$u(k, \ell) = f(k - \ell) + (\tfrac{2}{5})^k g(k - \ell) + \tfrac{1}{18}(3k - 4)k, \quad (5.212)$$

where f and g are arbitrary functions of $k - \ell$. Likewise, eliminating $u(k, \ell)$ gives

$$(E_1E_2 - 1)(5E_1E_2 - 2)v(k, \ell) = 2k + 1, \quad (5.213)$$

which can be solved to give

$$v(k, \ell) = F(k - \ell) + (\tfrac{2}{5})^k G(k - \ell) + \tfrac{1}{18}(3k - 1)k, \quad (5.214)$$

where F and G are arbitrary functions of $k - \ell$. The substitution of equations (5.212) and (5.214) into equation (5.210) allows the determination of $F(k - \ell)$ and $G(k - \ell)$ in terms of $f(k - \ell)$ and $g(k - \ell)$. Our final result is that the general solution to equations (5.210) is

$$\begin{aligned} u(k, \ell) &= f(k-\ell) + (2/5)^k g(k-\ell) + 1/18(3k-4)k, \\ v(k, \ell) &= f(k-\ell) - 2(2/5)^k g(k-\ell) + 1/18(3k-1)k. \end{aligned} \tag{5.215}$$

6
NONLINEAR DIFFERENCE EQUATIONS

6.1. INTRODUCTION

No general techniques exist for the solution of nonlinear difference equations. The purpose of this chapter is to present a number of methods which can be applied to obtain solutions to particular classes of nonlinear difference equations. The basic procedure will be to construct a nonlinear transformation such that applying it to the original equation leads to a new difference equation that is linear. For the most part, the nonlinear equations to be considered will be first order. It should be pointed out that nonlinear difference equations of degree higher than one may have more than one solution.

6.2. HOMOGENEOUS EQUATIONS

An equation homogeneous in y_k can be expressed in the following form:

$$f\left(\frac{y_{k+1}}{y_k}, k\right) = 0. \tag{6.1}$$

If the nonlinear function f is a polynomial function of y_{k+1}/y_k, then equation (6.1) can be written as

$$\prod_{i=1}^{n} [z_k - A_i(k)] = 0, \tag{6.2}$$

where $z_k = y_{k+1}/y_k$, $A_i(k)$ is a known function of k, and n is the order of the polynomial function of z_k. The solutions to each of the linear equations

$$z_k - A_i(k) = 0, \tag{6.3}$$

$$y_{k+1} - A_i(k)y_k = 0 \tag{6.4}$$

provide a solution to equation (6.1).

The following two examples illustrate the use of this technique.

6.2.1. Example A

Consider the equation

$$y_{k+1}^2 - 4y_{k+1}y_k - 5y_k^2 = 0. \tag{6.5}$$

Making the substitution $z_k = y_{k+1}/y_k$ gives

$$z_k^2 - 4z_k - 5 = (z_k - 5)(z_k + 1) = 0. \tag{6.6}$$

Therefore,

$$y_{k+1} - 5y_k = 0 \quad \text{or} \quad y_{k+1} + y_k = 0, \tag{6.7}$$

and, consequently,

$$y_k = C5^k \quad \text{or} \quad y_k = C(-1)^k. \tag{6.8}$$

6.2.2. Example B

The homogeneous equation

$$y_{k+1}^2 + (3-k)y_{k+1}y_k - 3ky_k^2 = 0 \tag{6.9}$$

can be written

$$(y_{k+1} + 3y_k)(y_{k+1} - ky_k) = 0. \tag{6.10}$$

Therefore, either

$$y_{k+1} + 3y_k = 0 \quad \text{or} \quad y_{k+1} - ky_k = 0, \tag{6.11}$$

and the solution to equation (6.9) is

$$y_k = C(-3)^k \quad \text{or} \quad y_k = C(k-1)!. \tag{6.12}$$

6.3. RICCATI EQUATIONS

The following nonlinear class of difference equations are of Riccati type:

$$P(k)y_{k+1}y_k + Q(k)y_{k+1} + R(k)y_k = 0, \tag{6.13}$$

where $P(k)$, $Q(k)$, and $R(k)$ are functions of k. The substitution

$$x_k = 1/y_k \tag{6.14}$$

gives the first-order, linear difference equation

$$R(k)x_{k+1} + Q(k)x_k + P(k) = 0, \tag{6.15}$$

which can be solved by using standard methods.

Equation (6.13) can be generalized to the form

$$P(k)y_{k+1}y_k + Q(k)y_{k+1} + R(k)y_k = S_k, \tag{6.16}$$

where again $P(k)$, $Q(k)$, $R(k)$, and $S(k)$ are functions of k. Division by $P(k)$ and shifting the index k gives

$$y_k y_{k-1} + A(k)y_k + B(k)y_{k-1} = C(k), \tag{6.17}$$

where

$$\begin{aligned} A(k) &\equiv \frac{Q(k-1)}{P(k-1)}, \\ B(k) &\equiv \frac{R(k-1)}{P(k-1)}, \\ C(k) &\equiv \frac{S(k-1)}{P(k-1)}. \end{aligned} \tag{6.18}$$

We will now show that the transformation

$$y_k = \frac{x_k - B(k)x_{k+1}}{x_{k+1}} \tag{6.19}$$

reduces equation (6.17) to a linear equation of second order.

The substitution of equation (6.19) into equation (6.17) gives

$$\begin{aligned} [x_k - B(k)x_{k+1}][x_{k-1} - B(k-1)x_k] + A(k)[x_k - B(k)x_{k+1}]x_k \\ + B(k)[x_{k-1} - B(k-1)x_k]x_{k+1} = C(k)x_{k+1}x_k, \end{aligned} \tag{6.20}$$

which can be simplied to the expression

$$x_k\{x_{k-1} + [A(k) - B(k-1)]x_k - [A(k)B(k) + C(k)]x_{k+1}\} = 0. \tag{6.21}$$

Thus, the required equation for x_k is

$$[A(k)B(k)+C(k)]x_{k+1}-[A(k)-B(k-1)]x_k-x_{k-1}=0, \quad (6.22)$$

which is linear and of second order.

At this point, several comments need to be made. First, equation (6.17) is of first order, while equation (6.22) is of second order. Therefore, we know that the general solutions to these two equations contain, respectively, one and two arbitrary constants. Consequently, one might conclude that the solution x_k obtained from equation (6.22) would lead to y_k having two arbitrary constants through the transformation of equation (6.19). However, an easy calculation shows that this is not the case and y_k depends on only one arbitrary constant. The proof proceeds as follows: Let the general solution to equation (6.22) be written

$$x_k = D_1 v_k + D_2 w_k, \quad (6.23)$$

where v_k and w_k are two linearly independent solutions and D_1 and D_2 are arbitrary constants. The substitution of equation (6.23) into equation (6.19) gives

$$y_k = \frac{[v_k - B(k)v_{k+1}]D_1 + [w_k - B(k)w_{k+1}]D_2}{D_1 v_{k+1} + D_2 w_{k+1}}, \quad (6.24)$$

or

$$y_k = \frac{[v_k - B(k)v_{k+1}] + [w_k - B(k)w_{k+1}]\overline{D}}{v_{k+1} + \overline{D}w_{k+1}}, \quad (6.25)$$

where $\overline{D} = D_2/D_1$. Therefore, we conclude that y_k depends on only one arbitrary constant.

Finally, it is easy to verify that the transformation

$$y_k = \frac{x_{k+1} - A(k+1)x_k}{x_k} \quad (6.26)$$

reduces equation (6.17) to the following second-order linear equation:

$$x_{k+1} - [A(k+1) - B(k+1)x_k - [A(k)B(k) + C(k)]x_{k-1} = 0. \quad (6.27)$$

6.3.1. Example A

Consider

$$y_{k+1}y_k + ay_{k+1} + by_k = 0, \tag{6.28}$$

where a, b, and c are constants. With $x_k = 1/y_k$, we obtain

$$bx_{k+1} + ax_k + 1 = 0, \tag{6.29}$$

which is a first-order linear inhomogeneous equation with constant coefficients. This equation has the solution

$$x_k = D\left(-\frac{a}{b}\right)^k - \frac{1}{a+b}, \qquad \text{if } a \neq -b, \tag{6.30}$$

$$x_k = D - \frac{1}{b}k, \qquad \text{if } a = -b, \tag{6.31}$$

where D is an arbitrary constant. The substitution of either of these results into $y_k = 1/x_k$ gives the general solution to equation (6.28).

6.3.2. Example B

The equation

$$y_{k+1}y_k + ay_{k+1} + by_k = c, \tag{6.32}$$

where a, b, and c are constants, is of the form given by equation (6.17). Thus, the substitution

$$y_k = x_k/x_{k+1} - b \tag{6.33}$$

transforms this equation to the form

$$(ab + c)x_{k+1} - (a - b)x_k - x_{k-1} = 0. \tag{6.34}$$

The characteristic equation for equation (6.34) is

$$(ab + c)r^2 - (a - b)r - 1 = 0. \tag{6.35}$$

If we denote the two roots by r_1 and r_2, then the solution to equation (6.34) can be written

$$x_k = D_1 r_1{}^k + D_2 r_2{}^k, \tag{6.36}$$

where D_1 and D_2 are arbitrary constants. Substituting equation (6.36) into equation (6.33) gives

$$y_k = \frac{(1 - br_1) + (1 - br_2)\overline{D}(r_2/r_1)^k}{r_1 + \overline{D}(r_2/r_1)^{k+1}}, \tag{6.37}$$

where $\overline{D}$ is an arbitrary constant. This is the general solution of equation (6.32).

6.3.3. Example C

The equation

$$y_{k+1} y_k - 2y_k = -2 \tag{6.38}$$

can be transformed into the linear equation

$$x_{k+2} - 2x_{k+1} + 2x_k = 0 \tag{6.39}$$

by means of the substitution

$$y_k = x_{k+1}/x_k. \tag{6.40}$$

The characteristic equation for equation (6.39) is

$$r^2 - 2r + 2 = 0, \tag{6.41}$$

and its two complex conjugate roots are

$$r_{1,2} = 1 \pm i = \sqrt{2}\, e^{\pm i\pi/4}. \tag{6.42}$$

Therefore, the general solution of equation (6.39) is

$$x_k = 2^{k/2}[D_1 \cos(\pi k/4) + D_2 \sin(\pi k/4)], \tag{6.43}$$

and

$$y_k = \sqrt{2}\,\frac{D_1 \cos[\pi(k+1)/4] + D_2 \sin[\pi(k+1)/4]}{D_1 \cos(\pi k/4) + D_2 \sin(\pi k/4)}. \tag{6.44}$$

Now, define the constant α such that in the interval $-\pi/2 < \alpha \leq \pi/2$,

$$\tan \alpha = D_2/D_1. \tag{6.45}$$

With this result, equation (6.44) becomes

$$y_k = \frac{\sqrt{2}\cos[(\pi/4)(k+1) - \alpha]}{\cos(\pi k/4 - \alpha)}, \tag{6.46}$$

or

$$y_k = 1 - \tan(\pi k/4 - \alpha). \tag{6.47}$$

This is the general solution to equation (6.38).

Note that since $\tan(\theta + \pi) = \tan \theta$, the solution has period 4; i.e., for given constant α, equation (6.47) takes on only four values; they are

$$\begin{aligned} y_0 &= 1 - \tan(-\alpha), \\ y_1 &= 1 - \tan(\pi/4 - \alpha), \\ y_2 &= 1 - \tan(\pi/2 - \alpha), \\ y_3 &= 1 - \tan(3\pi/4 - \alpha). \end{aligned} \tag{6.48}$$

An easy calculation shows that $y_0 = y_4$.

6.4. CLAIRAUT'S EQUATION

The following method often allows the determination of general solutions to first-order nonlinear difference equations.

Consider a nonlinear first-order difference equation having the form

$$f(k, y_k, \Delta y_k) = 0. \tag{6.49}$$

If we let $x_k = \Delta y_k$ and apply the Δ operator to equation (6.49), then the following result is obtained:

$$\Delta f = \phi(k, y_k, x_k, \Delta x_k) = 0. \tag{6.50}$$

If equation (6.50) is independent of y_k and if we can solve for x_k, we would obtain a relation of the form

$$g(k, x_k, A) = 0, \tag{6.51}$$

where A is an arbitrary constant. Consequently, the elimination of x_k between equations (6.49) and (6.51) will give a solution of equation (6.49).

As an application of this method consider the Clairaut difference equation

$$y_k = k\Delta y_k + f(\Delta y_k), \tag{6.52}$$

where f is a nonlinear function. Therefore,

$$y_k = kx_k + f(x_k) \tag{6.53}$$

and

$$\Delta y_k = x_k = (k+1)x_{k+1} - kx_k + f(x_{k+1}) - f(x_k). \tag{6.54}$$

Using the fact that

$$\begin{aligned} (k+1)x_{k+1} &= (k+1)\Delta x_k + (k+1)x_k, \\ x_{k+1} &= \Delta x_k + x_k, \end{aligned} \tag{6.55}$$

we obtain from equation (6.54)

$$(k+1)\Delta x_k + f(x_k + \Delta x_k) - f(x_k) = 0. \tag{6.56}$$

This last equation implies either

$$\Delta x_k = 0 \tag{6.57}$$

or

$$k + 1 + \frac{f(x_k + \Delta x_k) - f(x_k)}{\Delta x_k} = 0. \tag{6.58}$$

The first possibility, i.e., $\Delta x_k = 0$, gives

$$x_k = c = \text{constant}, \tag{6.59}$$

and from equation (6.53) the solution

$$y_k = ck + f(c). \tag{6.60}$$

The second possibility, equation (6.58), may lead to a second solution.

6.4.1. Example

Let the function f be equal to $(\Delta y_k)^2$; therefore, equation (6.52) becomes

$$y_k = kx_k + x_k{}^2, \tag{6.61}$$

where we have substituted $x_k = \Delta y_k$. Operating with Δ gives

$$(k+1)\Delta x_k + 2x_k \Delta x_k + (\Delta x_k)^2 = 0. \tag{6.62}$$

Thus, we conclude that either

$$\Delta x_k = 0 \quad \text{and} \quad y_k = ck + c^2, \tag{6.63}$$

or

$$\Delta x_k + 2x_k + k + 1 = x_{k+1} + x_k + k + 1 = 0. \tag{6.64}$$

The solution to the last equation is

$$x_k = c(-1)^k - \tfrac{1}{2}k - \tfrac{1}{4}, \tag{6.65}$$

which gives for equation (6.61) the second solution

$$y_k = [c(-1)^k - \tfrac{1}{4}]^2 - \tfrac{1}{4}k^2. \tag{6.66}$$

6.5. NONLINEAR TRANSFORMATIONS, MISCELLANEOUS FORMS

Consider the special class of nonlinear difference equations

$$(y_{k+n})^{\gamma_1}(y_{k+n-1})^{\gamma_2} \cdots (y_k)^{\gamma_{n+1}} = f(k), \tag{6.67}$$

where the γ_i are constants and $f(k)$ is a given function. This nth-order, nonlinear equation can be transformed into an nth-order, linear equation. To see this, take the logarithm of equation (6.67):

$$\gamma_1 \log y_{k+n} + \gamma_2 \log y_{k+n-1} + \cdots + \gamma_{n+1} \log y_k = \log f(k), \tag{6.68}$$

and define

$$x_k = \log y_k \quad \text{and} \quad g(k) = \log f(k). \tag{6.69}$$

Thus, x_k satisfies the following linear, inhomogeneous nth-order equation with constant coefficients:

$$\gamma_1 x_{k+n} + \gamma_2 x_{k+n-1} + \cdots + \gamma_{n+1} x_k = g(k). \tag{6.70}$$

This last equation can be solved by standard methods.

Note that if the solution of the difference equation

$$h(k, y_k, y_{k+1}, \ldots, y_{k+n}) = 0 \tag{6.71}$$

is known, then the solution of the difference equation

$$h[k, f(x_k), f(x_{k+1}), \ldots, f(x_{k+n})] = 0, \tag{6.72}$$

where f is a given function, is also known, since the last equation can be transformed into the first by the substitution

$$y_{k+i} = f(x_{k+i}), \qquad i = 0, 1, \ldots, n. \tag{6.73}$$

The following examples show how these techniques can be used.

6.5.1. Example A

Consider the nonlinear equation

$$y_{k+2} = (y_{k+1})^2 / y_k. \tag{6.74}$$

If we set $x_k = \log y_k$, then x_k satisfies the equation

$$x_{k+2} - 2x_{k+1} + x_k = 0, \tag{6.75}$$

the solution of which is

$$x_k = c_1 + c_2 k. \tag{6.76}$$

Therefore, a general solution of equation (6.74) is

$$y_k = e^{c_1 + c_2 k}, \tag{6.77}$$

or

$$y_k = c_3 e^{c_2 k}, \qquad c_3 > 0. \tag{6.78}$$

6.5.2. Example B

The equation

$$(k+1)y_{k+1}^2 - k y_k^2 = 0 \tag{6.79}$$

can be transformed to the linear form

$$x_{k+1} - x_k = 0 \tag{6.80}$$

by means of the substitution $x_k = k y_k^2$. Since $x_k = c$ is the solution of equation (6.80), we obtain

$$k y_k^2 = c \tag{6.81}$$

and

$$y_k = +\sqrt{c/k} \quad \text{or} \quad y_k = -\sqrt{c/k}, \tag{6.82}$$

as two solutions of equation (6.79).

6.5.3. Example C

If in the equation

$$(y_{k+2})^2 - 4(y_{k+1})^2 + 3(y_k)^2 = k \tag{6.83}$$

the substitution $x_k = y_k^2$ is made, then the following result is obtained:

$$x_{k+2} - 4x_{k+1} + 3x_k = k. \tag{6.84}$$

The solution to the latter equation is

$$x_k = c_1 + c_2 3^k - \tfrac{1}{4} k^2. \tag{6.85}$$

Thus,

$$y_k{}^2 = c_1 + c_2 3^k - \tfrac{1}{4} k^2$$

and equation (6.83) has the two solutions

$$y_k = +(c_1 + c_2 3^k - \tfrac{1}{4} k^2)^{1/2} \tag{6.86a}$$

or

$$y_k = -(c_1 + c_2 3^k - \tfrac{1}{4} k^2)^{1/2}. \tag{6.86b}$$

6.5.4. Example D

Consider the equation

$$\sqrt{y_{k+1}} = k\sqrt{y_k}\,. \tag{6.87}$$

Letting $x_k = \sqrt{y_k}$ gives

$$x_{k+1} = kx_k, \tag{6.88}$$

the solution of which is

$$x_k = c(k-1)!. \tag{6.89}$$

Therefore,

$$y_k = [c(k-1)!]^2. \tag{6.90}$$

We can obtain this solution to equation (6.87) by a second method. Taking the logarithm of both sides of equation (6.87) gives

$$z_{k+1} - z_k = \log k, \tag{6.91}$$

where $z_k = \tfrac{1}{2}(\log y_k)$. This last equation is of the form

$$z_{k+1} - p_k z_k = R_k \tag{6.92}$$

and can be solved by using the techniques of Chapter 1. Its solution is

$$z_k = \log\,[c(k-1)!], \tag{6.93}$$

and, consequently, the result given by equation (6.90) is obtained.

6.5.5. Example E

The nonlinear equation

$$y_{k+1} = (1 + y_k^{1/3})^3, \tag{6.94}$$

under the substitution $x_k = y_k^{1/3}$, becomes

$$x_{k+1} = x_k + 1, \tag{6.95}$$

the solution of which is

$$x_k = c + k. \tag{6.96}$$

Therefore, a solution to equation (6.94) is

$$y_k = (c + k)^3. \tag{6.97}$$

6.5.6. Example F

Consider the equation

$$y_{k+1} y_k y_{k-1} = A^2 (y_{k+1} + y_k + y_{k-1}), \tag{6.98}$$

where A is a constant. If we make use of the relationship

$$\tan(\theta_1 + \theta_2) = \frac{\tan\theta_1 + \tan\theta_2}{1 - \tan\theta_1 \tan\theta_2}, \tag{6.99}$$

and set

$$y_k = A \tan x_k, \tag{6.100}$$

then equation (6.98) becomes

$$\tan(x_{k+1} + x_k + x_{k-1}) = 0. \tag{6.101}$$

This last equation has the complete solution

$$x_{k+1} + x_k + x_{k-1} = n\pi, \tag{6.102}$$

where n is an integer.

The two roots to the characteristic equation corresponding to equation (6.102) are

$$r_1 = e^{i\phi}, \qquad r_2 = e^{-i\phi}, \qquad \phi = 2\pi/3. \tag{6.103}$$

Therefore,

$$x_k = c_1 \cos(2k\pi/3) + c_2 \sin(2k\pi/3) + n\pi/3, \tag{6.104}$$

and

$$y_k = A \tan[c_1 \cos(2k\pi/3) + c_2(2k\pi/3) + n\pi/3]. \tag{6.105}$$

Note that up to factors of $m\pi$, where m is an integer, equation (6.105) shows that there are three classes of solutions, corresponding to the "phase," $n\pi/3$, being equal to 0, $\pi/3$, or $2\pi/3$.

6.5.7. Example G

The substitution

$$y_k = \tfrac{1}{2}(1 - x_k) \tag{6.106}$$

transforms the equation

$$y_{k+1} = 2y_k(1 - y_k) \tag{6.107}$$

into

$$x_{k+1} = x_k{}^2, \tag{6.108}$$

the solution of which is

$$x_k = e^{c2^k}. \tag{6.109}$$

Therefore,

$$y_k = \tfrac{1}{2}(1 - e^{c2^k}). \tag{6.110}$$

6.5.8. Example H

The equation

$$y_{k+1} = 2y_k\sqrt{1 - y_k{}^2} \tag{6.111}$$

can be transformed into a simpler form by letting $y_k = \sin x_k$; this gives

$$\sin x_{k+1} = \sin 2x_k, \tag{6.112}$$

the solution of which is

$$x_{k+1} = (-1)^n 2x_k + n\pi, \tag{6.113}$$

where n is an integer. There are two cases to consider.

Let $n = 2m$ be an even integer. Therefore,

$$x_{k+1} - 2x_k = 2m\pi \tag{6.114}$$

and

$$x_k = c2^k - 2m\pi. \tag{6.115}$$

Consequently,

$$y_k = \sin(c2^k - 2m\pi) = \sin(c2^k). \tag{6.116}$$

Let $n = 2m + 1$ be an odd integer. In this instance, we have

$$x_{k+1} + 2x_k = (2m + 1)\pi, \tag{6.117}$$

$$x_k = c(-2)^k + \tfrac{1}{3}(2m + 1)\pi, \tag{6.118}$$

and

$$y_k = \sin[c(-2)^k + \tfrac{1}{3}(2m + 1)\pi]. \tag{6.119}$$

6.5.9. Example I

Let N be a positive number and consider the following nonlinear difference equation:

$$y_{k+1} = \tfrac{1}{2}(y_k - N/y_k). \tag{6.120}$$

Using the fact that

$$\cot(2\theta) = \frac{\cot^2\theta - 1}{2\cot\theta}, \tag{6.121}$$

we can make the substitution $y_k = \sqrt{N} \cot x_k$, reducing equation (6.120) to the form

$$\cot x_{k+1} = \cot(2x_k). \tag{6.122}$$

A solution of this latter equation is

$$x_{k+1} = 2x_k. \tag{6.123}$$

Therefore,

$$x_k = c2^k \tag{6.124}$$

and

$$y_k = \sqrt{N} \cot(c2^k). \tag{6.125}$$

6.6. PARTIAL DIFFERENCE EQUATIONS

The situation regarding obtaining solutions to nonlinear partial difference equations is similar to that for ordinary difference equations; namely, except under special conditions, no general techniques exist for finding solutions. Nonlinear transformation can often be used to reduce certain classes of nonlinear partial difference equations to linear forms which can then be solved by using the methods presented in Chapter 5.

A particular nonlinear equation which can be immediately solved is *Clairaut's extended form,* which is given by the following expression:

$$z(k, l) = k\Delta_1 z(k, l) + l\Delta_2 z(k, l) + f[\Delta_1 z(k, l), \Delta_2 z(k, l)], \tag{6.126}$$

where

$$\begin{aligned} \Delta_1 z(k, l) &= z(k+1, l) - z(k, l), \\ \Delta_2 z(k, l) &= z(k, l+1) - z(k, l). \end{aligned} \tag{6.127}$$

If c_1 and c_2 are arbitrary constants, then a solution of equation (6.126) is

$$z(k, l) = c_1 k + c_2 l + f(c_1, c_2). \tag{6.128}$$

The *Riccati extended form* is a nonlinear system of equations which can be solved exactly. The equations of interest are

$$u(k+1, l+1) = \frac{au(k, l) + bv(k, l) + c}{pu(k, l) + qv(k,l) + r},$$
$$v(k+1, l+1) = \frac{Au(k, l) + Bv(k, l) + C}{pu(k, l) + qv(k, l) + r}, \tag{6.129}$$

where a, b, c, A, B, C, p, q, and r are constants. We will now indicate how the required solutions can be found.

From equations (6.129), it follows that if λ, μ, and ν are undetermined multipliers, then the following relations hold:

$$\frac{u(k+1, l+1)}{au(k, l) + bv(k, l) + c} = \frac{v(k+1, l+1)}{Au(k, l) + Bv(k, l) + C}$$
$$= \frac{1}{pu(k, l) + qv(k, l) + r} \tag{6.130}$$
$$= \frac{\lambda u(k+1, l+1) + \mu v(k+1, l+1) + \nu}{[(a\lambda + A\mu + p\nu)u(k, l) + (b\lambda + B\mu + q\nu)v(k, l) + (c\lambda + C\mu + r\nu)]}.$$

Let λ, μ, and ν be chosen such that

$$a\lambda + A\mu + p\nu = h\lambda,$$
$$b\lambda + B\mu + q\nu = h\mu, \tag{6.131}$$
$$c\lambda + C\mu + r\nu = h\nu,$$

where h is, for the moment, an unknown constant. The condition that λ, μ, and ν not be zero is

$$\begin{vmatrix} a-h & A & p \\ b & B-h & q \\ c & C & r-h \end{vmatrix} = 0. \tag{6.132}$$

Note that equation (6.132) is a cubic equation in h and, therefore, has three solutions, h_1, h_2, and h_3. Each of these solutions allows us to determine from equation (6.131) corresponding values of λ, μ, and ν, i.e.,

$$h_1 \rightarrow (\lambda_1, \mu_1, \nu_1),$$
$$h_2 \rightarrow (\lambda_2, \mu_2, \nu_2), \tag{6.133}$$
$$h_3 \rightarrow (\lambda_3, \mu_3, \nu_3).$$

Now define $P_1(k, l)$, $P_2(k, l)$, and $P_3(k, l)$ to be

$$P_i(k, l) = \lambda_i u(k, l) + \mu_i v(k, l) + \nu_i, \qquad i = 1, 2, 3. \tag{6.134}$$

Therefore, equations (6.130) can be written

$$\frac{P_1(k+1, l+1)}{h_1 P_1(k, l)} = \frac{P_2(k+1, l+1)}{h_2 P_2(k, l)} = \frac{P_3(k+1, l+1)}{h_3 P_3(k, l)}. \tag{6.135}$$

From these latter equations, the following relations hold:

$$\begin{aligned} \frac{P_1(k+1, l+1)}{P_3(k+1, l+1)} &= \frac{h_1}{h_3}\frac{P_1(k, l)}{P_3(k, l)}, \\ \frac{P_2(k+1, l+1)}{P_3(k+1, l+1)} &= \frac{h_2}{h_3}\frac{P_2(k, l)}{P_3(k, l)}. \end{aligned} \tag{6.136}$$

Now, the equations (6.136) are of the form

$$z(k+1, l+1) = \beta z(k, l), \tag{6.137}$$

where β is a constant. Such equations have been solved in Chapter 5. Using this result, we obtain

$$\begin{aligned} \frac{P_1(k, l)}{P_3(k, l)} &= \left(\frac{h_1}{h_3}\right)^k \frac{\phi_1(k-l)}{\phi_3(k-l)} \equiv \psi_1(k, l), \\ \frac{P_2(k, l)}{P_3(k, l)} &= \left(\frac{h_2}{h_3}\right)^k \frac{\phi_2(k-l)}{\phi_3(k-l)} \equiv \psi_2(k, l), \end{aligned} \tag{6.138}$$

where ϕ_1, ϕ_2, and ϕ_3 are arbitrary functions of $k - l$. Substituting equations (6.134) into equations (6.138) and simplifying the resulting expressions gives

$$\begin{aligned} (\lambda_1 - \lambda_3\psi_1)u(k, l) + (\mu_1 - \mu_3\psi_1)v(k, l) &= \nu_3\psi_1 - \nu_1, \\ (\lambda_2 - \lambda_3\psi_2)u(k, l) + (\mu_2 - \mu_3\psi_2)v(k, l) &= \nu_3\psi_2 - \nu_2. \end{aligned} \tag{6.139}$$

These equations can be solved for $u(k, l)$ and $v(k, l)$ in terms of the constants $(\lambda_1, \lambda_2, \lambda_3, \mu_1, \mu_2, \mu_3)$ and the functions $\psi_1(k, l)$ and $\psi_2(k, l)$.

6.6.1. Example A

The nonlinear equation

$$z(k+1, l) = [z(k, l+1)]^\gamma, \tag{6.140}$$

where γ is a constant, can be transformed to the linear equation

$$w(k+1, l) = \gamma w(k, l+1) \tag{6.141}$$

by means of the substitution

$$w(k, l) = \log z(k,l). \tag{6.142}$$

Equation (6.141) has the solution

$$w(k, l) = \gamma^k f(k+l), \tag{6.143}$$

where f is an arbitrary function of $(k + l)$. Therefore, equation (6.140) has the solution

$$z(k, l) = \exp[w(k, l)] = \exp[\gamma^k f(k+l)]. \tag{6.144}$$

6.6.2. Example B

The equation

$$z(k, l) = k\Delta_1 z(k, l) + l\Delta_2 z(k, l) + (\Delta_1 z)^2 \cos[\Delta_2 z(k, l)] \tag{6.145}$$

has the solution

$$z(k, l) = Ak + Bl + A^2 \cos B, \tag{6.146}$$

where A and B are arbitrary constants.

APPENDIX
USEFUL MATHEMATICAL RELATIONS

A. ALGEBRAIC RELATIONS

A.1. Binomial Theorem

$$(a + x)^n = a^n + \binom{n}{n-1} a^{n-1}x + \binom{n}{n-2} a^{n-2}x^2 + \cdots + \binom{n}{1} ax^{n-1} + x^n,$$

where the binomial coefficients are

$$\binom{n}{k} = \frac{n!}{(n-k)!k!} = \frac{n(n-1)\cdots(n-k+1)}{1 \cdot 2 \cdot 3 \cdot \cdots \cdot k}.$$

A.2. Factors and Expansions

$$(a \pm b)^2 = a^2 \pm 2ab + b^2$$

$$(a \pm b)^3 = a^3 \pm 3a^2b + 3ab^2 \pm b^3$$

$$(a + b + c)^2 = a^2 + b^2 + c^2 + 2(ab + ac + bc)$$

$$(a + b + c)^3 = a^3 + b^3 + c^3 + 3a^2(b + c) + 3b^2(a + c) + 3c^2(a + b) + 6abc$$

$$a^2 - b^2 = (a - b)(a + b)$$

$$a^2 + b^2 = (a + ib)(a - ib), \qquad i = \sqrt{-1}$$

$$a^3 - b^3 = (a - b)(a^2 + ab + b^2)$$

$$a^3 + b^3 = (a + b)(a^2 - ab + b^2)$$

B. TRIGONOMETRIC RELATIONS

B.1. Exponential Definitions of Trigonometric Functions

$$\sin\theta = \frac{e^{i\theta} - e^{-i\theta}}{2i}$$

$$\cos\theta = \frac{e^{i\theta} + e^{-i\theta}}{2}, \qquad i = \sqrt{-1}$$

B.2. Functions of Sums of Angles

$$\sin(\theta_1 \pm \theta_2) = \sin\theta_1 \cos\theta_2 \pm \cos\theta_1 \sin\theta_2$$

$$\cos(\theta_2 \pm \theta_2) = \cos\theta_1 \cos\theta_2 \mp \sin\theta_1 \sin\theta_2$$

B.3. Powers of Trigonometric Functions

$$\sin^2\theta = \tfrac{1}{2}(1 - \cos 2\theta)$$

$$\cos^2\theta = \tfrac{1}{2}(1 + \cos 2\theta)$$

$$\sin^3\theta = \tfrac{1}{4}(3\sin\theta - \sin 3\theta)$$

$$\cos^3\theta = \tfrac{1}{4}(3\cos\theta + \cos 3\theta)$$

$$\sin^4\theta = \tfrac{1}{8}(3 - 4\cos 2\theta + \cos 4\theta)$$

$$\cos^4\theta = \tfrac{1}{8}(3 + 4\cos 2\theta + \cos 4\theta)$$

$$\sin^5\theta = \tfrac{1}{16}(10\sin\theta - 5\sin 3\theta + \sin 5\theta)$$

$$\cos^5\theta = \tfrac{1}{16}(10\cos\theta + 5\cos 3\theta + \cos 5\theta)$$

B.4. Other Trigonometric Relations

$$\sin\theta_1 \pm \sin\theta_2 = 2\sin\left(\frac{\theta_1 \pm \theta_2}{2}\right)\cos\left(\frac{\theta_1 \mp \theta_2}{2}\right)$$

$$\cos\theta_1 + \cos\theta_2 = 2\cos\left(\frac{\theta_1 + \theta_2}{2}\right)\cos\left(\frac{\theta_1 - \theta_2}{2}\right)$$

$$\cos\theta_1 - \cos\theta_2 = -2\sin\left(\frac{\theta_1 + \theta_2}{2}\right)\sin\left(\frac{\theta_1 - \theta_2}{2}\right)$$

$$\sin\theta_1 \cos\theta_2 = \tfrac{1}{2}[\sin(\theta_1 + \theta_2) + \sin(\theta_1 - \theta_2)]$$

$$\cos\theta_1 \sin\theta_2 = \tfrac{1}{2}[\sin(\theta_1 + \theta_2) - \sin(\theta_1 - \theta_2)]$$

$$\cos\theta_1 \cos\theta_2 = \tfrac{1}{2}[\cos(\theta_1 + \theta_2) + \cos(\theta_1 - \theta_2)]$$

$$\sin\theta_1 \sin\theta_2 = \tfrac{1}{2}[\cos(\theta_1 + \theta_2) - \cos(\theta_1 - \theta_2)]$$

C. CONTINUOUS FUNCTIONS

C.1. Taylor's Theorems

If a function $f(x)$ has derivatives of all orders in a neighborhood of a point $x = a$, then

$$f(x) = f(a) + \frac{f'(a)}{1!}(x-a) + \frac{f''(a)}{2!}(x-a)^2 + \cdots + \frac{f^{(n)}(a)}{n!}(x-a)^n + \cdots .$$

If a function $f(x, y)$ has derivatives of all orders in the neighborhood of $x = a$ and $y = b$, then

$$f(x, y) = f(a, b) + \frac{\partial f(a, b)}{\partial x}(x-a) + \frac{\partial f(a, b)}{\partial y}(y-b) + \frac{1}{2!}\left(\frac{\partial^2 f(a, b)}{\partial x^2}(x-a)^2 + 2\frac{\partial^2 f(a, b)}{\partial x\,\partial y}(x-a)(y-b) + \frac{\partial^2 f(a, b)}{\partial y^2}(y-b)^2\right) + \cdots$$

C.2. Expansion of a Function of a Function

Let $f(y)$ be a function of y and let $y = g(x)$ be a function of x. If

$$y(x) = a_0 + a_1 x + a_2 x^2 + a_3 x^4 + \cdots ,$$

then

$$f[y(x)] = A_0 + A_1 x + A_2 x^2 + A_3 x^3 + \cdots ,$$

where

$$A_0 = f(a_0),$$

$$A_1 = a_1 f'(a_0),$$

$$A_2 = [a_1^2 f''(a_0) + 2a_2 f'(a_0)]/2!,$$

$$A_3 = [a_1^3 f'''(a_0) + 6a_1 a_2 f''(a_0) + 6a_3 f'(a_0)]/3!,$$

$$A_4 = [a_1^4 f''''(a_0) + 12a_1^2 a_2 f'''(a_0) + (24a_1 a_3 + 12a_2^2) f''(a_0) + 24a_4 f'(a_0)]/4!,$$

$$A_5 = [a_1^5 f'''''(a_0) + 20a_1^3 f''''(a_0) + 60a_1(a_1 a_3 + a_2^2) f'''(a_0) + 120(a_1 a_4 + a_2 a_3) f''(a_0) + 120a_5 f'(a_0)]/5!.$$

C.3. D'Alembert's Ratio Test for Convergence

The series

$$\sum_{k=1}^{\infty} u_k = u_1 + u_2 + u_3 + \cdots$$

is said to converge absolutely if the series

$$\sum_{k=1}^{\infty} |u_k| = |u_1| + |u_2| + |u_3| + \cdots$$

converges. If the original series converges and the series composed of the absolute values diverges, then the original series is said to converge conditionally. Every absolutely convergent series converges.

Assume that

$$\lim_{k\to\infty} \left| \frac{u_{k+1}}{u_k} \right| = r.$$

If $r < 1$, then the series converges absolutely. If $r > 1$, the series diverges. For $r = 1$, no conclusion can be reached.

C.4. Leibnitz's Rule for the *n*th Derivative of a Product

Let $u(x)$ and $v(x)$ be n-times differentiable functions of x. Then

$$\frac{d^n(uv)}{dx} = u\frac{d^n v}{dx^n} + \binom{n}{1}\frac{du}{dx}\frac{d^{n-1}v}{dx^{n-1}} + \binom{n}{2}\frac{d^2u}{dx^2}\frac{d^{n-2}v}{dx^{n-2}} + \cdots + v\frac{d^n u}{dx^{n-1}}.$$

C.5. Expansions of Selected Functions

$$(a+x)^n = \sum_{k=0}^{n} \binom{n}{k} a^{n-k}x^k$$

$$(1+x)^{-1} = 1 - x + x^2 - x^3 + \cdots = \sum_{k=0}^{\infty} (-1)^k x^k$$

$$(1+x)^{-2} = 1 - 2x + 3x^2 - 4x^3 + \cdots = \sum_{k=0}^{\infty} (-1)^k (k+1) x^k$$

$$\frac{x}{(1-x)^2} = x + 2x^2 + 3x^3 + \cdots = \sum_{k=1}^{\infty} kx^k$$

$$e^x = 1 + x + \frac{x^2}{2!} + \frac{x^3}{3!} + \cdots = \sum_{k=0}^{\infty} \frac{x^k}{k!}$$

D. PARTIAL EXPANSIONS

A function $f(x)$ is said to be rational if it is the ratio of two polynomials in x,

$$f(x) = \frac{b_0 x^m + b_1 x^{m-1} + \cdots + b_m}{x^n + a_1 x^{n-1} + \cdots + a_n}.$$

A rational function is said to be proper if the degree of the numerator polynomial is no greater than the degree of the denominator polynomial. It is strictly proper if the degree of the numerator polynomial is less than the degree of the denominator polynomial. The rational function is said to be reduced if the numerator and denominator polynomials have no common factors. Finally, the degree of a rational function is equal to the degree of the denominator polynomial when the function is reduced.

The coefficients of a partial-fraction expansion can be determined as follows. Assume $f(x)$ is a reduced, strictly proper rational function of degree n.

(i) Let the roots of the denominator polynomial be distinct. Then $f(x)$ has the partial-fraction expansion

$$f(x) = \frac{c_1}{x - x_1} + \frac{c_2}{x - x_2} + \cdots + \frac{c_n}{x - x_n},$$

where

$$c_r = \lim_{x \to x_r} (x - x_r) f(x), \qquad i = 1, 2, \ldots, n,$$

and x_r denotes the rth root.

(ii) Let x_i be a root of multiplicity m. For this case, the partial fraction of $f(x)$ includes terms of the form

$$\frac{c_{i1}}{x - x_i} + \frac{c_{i2}}{(x - x_i)^2} + \cdots + \frac{c_{im}}{(x - x_i)^m},$$

where

$$c_{ik} = \lim_{x \to x_i} \frac{1}{(m-k)!} \frac{d^{m-k}}{dx^{m-k}} [(x - x_i)^m f(x)], \qquad 1 \le k \le m.$$

E. THE VANDERMONDE DETERMINANT

The determinant

$$D = \begin{vmatrix} 1 & 1 & \cdots & 1 \\ \lambda_1 & \lambda_2 & \cdots & \lambda_n \\ \lambda_1^2 & \lambda_2^2 & \cdots & \lambda_n^2 \\ \cdot & \cdot & & \cdot \\ \cdot & \cdot & & \cdot \\ \cdot & \cdot & & \cdot \\ \lambda_1^{n-1} & \lambda_2^{n-1} & \cdots & \lambda_n^{n-1} \end{vmatrix}$$

is called a Vandermonde determinant. It has the expansion

$$D = \prod_{1 \le i < j \le n} (\lambda_j - \lambda_i).$$

F. SEQUENCES

A sequence is a function whose domain is the set of integers. Thus, for each integer k, the function f defines a value f_k. The sequence is represented by the symbol $\{f_k\}$, where f_k denotes the kth term.

In general, the members of a sequence may be real or complex numbers, vectors, functions, matrices, etc.

The following is a list of properties and definitions which are of importance in understanding and using sequences. We assume that the members of the sequence consist of real numbers.

(i) A real sequence $\{f_k\}$ is said to be bounded if a real constant M exists such that $|f_k| < M$ for all k. If no such constant M exists, the sequence is unbounded.

(ii) A sequence of real numbers $\{f_k\}$ is said to be monotone increasing if $f_{k+1} > f_k$ for all k. The sequence is said to be monotone decreasing if $f_{k+1} \le f_k$ for all k.

(iii) A real sequence $\{f_k\}$ is called a null sequence if, corresponding to any positive number ϵ, no matter how small, there exists a positive integer N, depending on ϵ, such that $|f_k| < \epsilon$ for all $k > N$. Thus for a real null sequence, all but a finite number of members must lie in the interval $(-\epsilon, \epsilon)$, i.e., $-\epsilon < f_k < \epsilon$.

(iv) Let $\{c_k\}$ be a bounded sequence and let $\{f_k\}$ be a null sequence; then $\{c_k f_k\}$ is a null sequence.

(v) Let $\{f_k\}$ be a real sequence. Let there exist a real number L such that $\{f_k - L\}$ is a null sequence. Then $\{f_k\}$ is said to converge to the number L, which is called the limit of the sequence. A real sequence is divergent if it is not convergent.

(vi) The sequence $\{f_k\}$, if convergent with limit L, is said to be damped oscillatory if each member of the sequence less than L is followed by some member (not necessarily the next one) greater than L. Likewise, each member of the sequence greater than L is followed by some member (again, not necessarily the next one) less than L.

(vii) The sequence $\{f_k\}$ is said to diverge to $+\infty$ if for any positive number M there exists a positive integer K, depending on M, such that $f_k > M$ for all $k > K$. (A similar definition holds for divergence to $-\infty$.)

(viii) Let the sequence $\{f_k\}$ be divergent, but one that does not diverge to either $+\infty$ or $-\infty$. Then $\{f_k\}$ is said to oscillate finitely if it is bounded and to oscillate infinitely if it is unbounded.

(ix) Let $\{f_k\}$ and $\{g_k\}$ be convergent sequences with respective limits L_1 and L_2. Then

$$\lim_{k\to\infty} (c_1 f_k + c_2 g_k) = c_1 L_1 + c_2 L_2,$$

$$\lim_{k\to\infty} f_k g_k = L_1 L_2,$$

$$\lim_{k\to\infty} \frac{f_k}{g_k} = \frac{L_1}{L_2}, \qquad \text{if } L_2 \neq 0,$$

c_1 and c_2 are constants.

G. GAMMA FUNCTION

We now define a function known as the gamma function, $\Gamma(x)$, where x is a real continuous variable. (With little additional work, the gamma function can be extended to be a function of a complex variable.) This function has the property that $\Gamma(k) = (k - 1)!$ for every positive integer k. Consequently, it can be regarded as a generalization of $k!$ to the situation where the value of the variable is not an integer.

The gamma function is defined in terms of an improper integral which cannot be evaluated in terms of elementary functions. Since this function plays a role of great significance in mathematical analysis and in application in various scientific fields, it has been studied in great detail and its values tabulated.

The following listing gives several of the important properties of the gamma function.

(i) The gamma function is defined by the integral

$$\Gamma(x) = \int_{0^+}^{\infty} e^{-t} t^{x-1}\, dt, \qquad 0 < x < \infty.$$

(ii) $\Gamma(x+1) = x\Gamma(x), \qquad 0 < x < \infty.$

(iii) $\Gamma(x+p) = (x+p-1)(x+p-2)\cdots x\Gamma(x), \qquad x > 0,\ p = 1, 2, 3, \ldots.$

(iv) $\Gamma(n+1) = n!, \qquad n = 0, 1, 2, \ldots.$

(v) $\Gamma(0^+) = +\infty.$

(vi) $\Gamma(x)$ is a continuous function for $x > 0$.

(vii) $\lim_{x \to 0^+} x\Gamma(x) = 1.$

(viii) For $n = 1, 2, 3, \ldots$, and $-n < x < -n+1$, we define

$$\Gamma(x) = \frac{\Gamma(x+n)}{x(x+1)\cdots(x+n-1)}.$$

(ix) Using these results $\Gamma(x)$ can now be defined for all x, except zero and the negative integers, i.e.,

$$\Gamma(x+1) = x\Gamma(x), \qquad x \neq 0, -1, -2, \ldots.$$

(x) From the results of (viii) and (ix), it is clear that it is only necessary to know the gamma function in an interval of length one; all other values can then be determined.

H. PRINCIPLE OF MATHEMATICAL INDUCTION

Assume that to each positive integer k there corresponds a statement $S(k)$ which is either true or false. Assume further (i) $S(1)$ is true and (ii) if $S(k)$ is true, then $S(k+1)$ is true for all k. Then $S(k)$ is true for all positive integers k.

NOTES AND REFERENCES

This section gives references that provide additional background material and/or extensions of particular topics that are discussed in the book. References to books in the bibliography are indicated by name of author(s) (year of publication). Complete information on references to books that do not appear in the bibliography and papers are given at the time that they are introduced.

CHAPTER 1

Section 1.4: The conditions for existence and uniqueness of the initial-value problem for implicit nonlinear difference equations are given in the technical report of Thomas K. Caughey (Jet Propulsion Laboratory Publication 79–50, California Institute of Technology, June 1, 1979). A unified treatment of the existence and uniqueness theorems for difference equations and systems of simultaneous difference equations in normal form (i.e., the equations can be solved with respect to the highest-order difference term) are presented in a paper by Selmo Tauber [*American Mathematical Monthly* **71,** 859 (1964).

Section 1.7: A detailed discussion of the factorial polynomials, the Stirling numbers, solution of the associated difference equations, and applications is presented in Sections 16 and 50–77 of Jordan (1956). See also Miller (1960), Sections 1.3–1.5.

CHAPTER 2

Section 2.3: An excellent introduction and detailed review of the properties and applications of Bernoulli numbers and polynomials is given in Miller (1960), Chapter 3. See also Milne-Thomson (1960), Chapter X.

Section 2.5: The following book provides an elementary introduction to continued fractions: Carl D. Olds, *Continued Fractions* (Random House, New York, 1963). A more advanced text which discusses the analytical aspects of continued fractions is Hubert S. Wall, *Analytic Theory of Continued Fractions* (Van Nostrand, New York, 1948).

Sections 2.7 and 2.8: The most detailed presentation, in a textbook, of geometrical and analytical expansion techniques to investigate the asymptotic behavior of the solutions of nonlinear first-order difference equations is that given in Levy and Lessman (1961), Chapter 5. The asymptotics of analytic difference equations is discussed in Immink (1981). Bender and Orszag (1978), Chapter 5, analyze the approximate solution of difference equations near ordinary and regular singular points of linear difference equations, the local behavior near an irregular singular point, and the local behavior of particular nonlinear difference equations. Exact solutions are determined for several nonlinear problems. The use of discrete mapping techniques to investigate complex physical systems is given by Devaney (1986) and Schuster (1984).

CHAPTER 3

Section 3.6: The books by Fort (1948), Chapters X, XIV, and XV, and Hildebrand (1965), Sections 1.10–1.16, treat Sturm–Liouville difference equations including characteristic or eigenvalue problems, symmetry and periodicity conditions, and discrete Fourier representations.

CHAPTER 4

Section 4.4: The results of this section are based on a paper by David Zeitlin [American Mathematical Monthly **68,** 369 (1961).

Section 4.7: Bishop (1975), Chapters V and VI, and Jury (1964) treat in detail both the theory and application of the z-transform.

CHAPTER 5

Section 5.5: A good presentation of various techniques for obtaining particular solutions to certain classes of inhomogeneous partial difference equations is given by Levy and Lessman (1961), Chapter 8. Jordan (1956), Chapter XII, treats partial difference equations in three and four independent variables.

CHAPTER 6

In addition to the references given under Sections 2.7 and 2.8 above, the book by Bernussou (1977) treats point mapping stability. The book of Vidal (1969) deals primarily with nonlinear discrete-time systems. A number of techniques are presented for determining approximate solutions. The stability of discrete-time systems is also discussed.

APPENDIX

Section C: Taylor's series and related matters are presented in detail with proofs in Angus E. Taylor and William Robert Mann, *Advanced Calculus* (John Wiley & Sons, New York, 1983); Sections 4.2, 7.5, and 19.1.

Section D: Excellent descriptions of the partial expansion of a rational function are given in Bishop (1975), Section 6.5, and Jordan (1956), Section 13.

Section E: A brief discussion of the properties of the Vandermonde determinant is given in Richard Bellman, *Introduction to Matrix Analysis* (McGraw Hill, New York, 1970), p. 193.

Section F: Good discussions of sequences are presented by Wilfred Kaplan, *Advanced Calculus* (Addison-Wesley, Reading, MA, 1952), Sections 6–2 and 6–3, and Kenkel (1974), Section 1.2.

Section G: The properties of the gamma functions, the beta function, and the evaluation of certain definite integrals related to the gamma function and Stirling's formula for the asymptotic expansion of *n!* are given in David V. Widder, *Advanced Calculus* (Prentice-Hall, Englewood Cliffs, NJ, 1961); Section 11.1.

BIBLIOGRAPHY

GENERAL THEORY

Paul M. Batchelder, *An Introduction to Linear Difference Equations* (Harvard University Press, Cambridge, 1927).

George Boole, *Calculus of Finite Differences,* 4th ed. (Chelsea, New York, 1958).

Louis Brand, *Differential and Difference Equations* (Wiley, New York, 1966).

Frank Chorlton, *Differential and Difference Equations* (Van Nostrand, London, 1965).

Edward J. Cogan and Robert Z. Norman, *Handbook of Calculus, Difference and Differential Equations* (Prentice-Hall, Englewood Cliffs, NJ, 1958).

Tomlinson Fort, *Finite Differences and Difference Equations in the Real Domain* (Clarendon Press, Oxford, 1948).

Charles Jordan, *Calculus of Finite Differences,* 3rd ed. (Chelsea, New York, 1965).

H. Levy and F. Lessman, *Finite Difference Equations* (Macmillan, New York, 1961).

Kenneth S. Miller, *An Introduction to the Calculus of Finite Differences and Difference Equations* (Holt, New York, 1960).

Kenneth S. Miller, *Linear Difference Equations* (W. A. Benjamin, New York, 1968).

Louis M. Milne-Thomson, *The Calculus of Finite Differences* (Macmillan, London, 1960).

Clarence H. Richardson, *An Introduction to the Calculus of Finite Differences* (Van Nostrand, New York, 1954).

Murray R. Spiegel, *Calculus of Finite Differences and Difference Equations* (McGraw-Hill, New York, 1971).

APPLICATIONS AND ADVANCED TOPICS

Rutherford Aris, *Discrete Dynamic Programming* (Blaisdell, New York, 1964).

Carl M. Bender and Steven A. Orszag, *Advanced Mathematical Methods for Scientists and Engineers* (McGraw-Hill, New York, 1978).

Jacques Bernussou, *Point Mapping Stability* (Pergamon, New York, 1977).

Albert B. Bishop, *Introduction to Discrete Linear Controls* (Academic, New York, 1975).

James A. Cadzow, *Discrete-Time Systems* (Prentice-Hall, Englewood Cliffs, NJ, 1973).

Carl F. Christ, *Econometric Models and Methods* (Wiley, New York, 1966).

Thomas F. Dernburg and Judith D. Dernburg, *Macroeconomic Analysis* (Addison-Wesley, Reading, MA, 1969).

Robert L. Devaney, *An Introduction to Chaotic Dynamical Systems* (Benjamin/Cummings, Menlo Park, CA, 1986).

George E. Forsythe and Wolfgang R. Wasow, *Finite-Difference Methods for Partial Differential Equations* (Wiley, New York, 1960).

A. O. Gel'fond, *Calculus of Finite Differences* (Hinduston, Delhi, India, 1971).

Sergei K. Godunov and V. S. Ryabenki, *Theory of Difference Schemes* (North-Holland, Amsterdam, 1964).

Samuel Goldberg, *Introduction to Difference Equations* (Wiley, New York, 1958).

Donald Greenspan, *Discrete Models* (Addison-Wesley, Reading, MA, 1973).

Donald Greenspan, *Arithmetic Applied Mathematics* (Pergamon, Oxford, 1980).

Peter Henrici, *Discrete Variable Methods in Ordinary Differential Equations* (Wiley, New York, 1962).

Peter Henrici, *Error Propagation for Finite Difference Methods* (Wiley, New York, 1963).

Francis B. Hildebrand, *Methods of Applied Mathematics,* 2nd ed. (Prentice-Hall, Englewood Cliffs, NJ, 1965).

Francis B. Hildebrand, *Finite-Difference Equations and Simulations* (Prentice-Hall, Englewood Cliffs, NJ, 1968).

Geertrui K. Immink, *Asymptotics of Analytic Difference Equations* (Springer-Verlag, New York, 1981).

Roy M. Johnson, *Theory and Applications of Linear Differential and Difference Equations* (Halsted, New York, 1984).

Eliahu I. Jury, *Sampled-Data Control systems* (Wiley, New York, 1958).

Eliahu I. Jury, *Theory and Application of the z-Transform* (Wiley, New York, 1964).

James L. Kenkel, *Dynamic Linear Economic Models* (Gordon and Breach, New York, 1974).

David G. Luenberger, *Introduction to Dynamic Systems* (Wiley, New York, 1979).

William E. Milne, *Numerical Calculus* (Princeton University Press, Princeton, NJ, 1949).

Thomas L. Saaty, *Modern Nonlinear Equations* (McGraw-Hill, New York, 1967).

Heinz Georg Schuster, *Deterministic Chaos* (Physik-Verlag, Weinheim, West Germany, 1984).

Vladimir Strejc, *State Space Theory of Discrete Linear Control* (Wiley-Interscience, New York, 1981).

Knut Sydsaeter, *Topics in Mathematical Analysis for Economists* (Academic, New York, 1981).

Pierre Vidal, *Non-Linear Sampled-Data Systems* (Gordon and Breach, New York, 1969).

Jet Wimp, *Computation and Recurrence Relations* (Pitman, Marshfield, MA, 1984).

INDEX